Active-Filter Cookbook

by

Don Lancaster

SAMS

A Division of Prentice Hall Computer Publishing
11711 North College, Carmel, Indiana 46032 USA

International Standard Book Number: 0-672-21168-8
Library of Congress Catalog Card Number: 74-33839

95   94   93   92   20   19   18   17

Interpretation of the printing code: the rightmost
double-digit number in the first column is the year of
the book's printing; the rightmost double-digit number
in the second column, the number of the book's
printing. For example, a printing code of 92-17 shows
that the seventeenth printing of the book occurred
in 1992.

Printed in the United States of America.

TK
7872
F5
L 28

Preface

An *active filter* is some combination of integrated-circuit operational amplifiers, resistors, and capacitors that does things that normally could be done only with expensive inductor-capacitor passive filter combinations. Active filters are versatile, low-cost items that are easy to design and easy to tune. They have gain and have a number of other benefits. Active filters are well suited for most subaudio, audio, and ultrasonic filtering or equalizing applications. Important areas of use for active filters include communications, electronic music, brainwave research, quadrature art, speech and hearing studies, telephony, psychedelic lighting, medical electronics, seismology, instrumentation, and many other areas.

This book is about active filters. It is user-oriented. It tells you everything you need to know to build active filters, and does so with an absolute minimum of math or obscure theory.

If you know nothing at all about active filters and simply need a frequency-selective circuit, this book will serve you as a catalog of "ripoff" circuits that are ready for immediate use—with math ranging from none at all to one or two simple multiplications.

If you are interested in the how and why of active filters, there is more-detailed information here that lets you do more-involved design work, optimizing things to your particular needs and perhaps using a simple handheld calculator for the actual final design effort.

Finally, if you are an active-filter specialist, you will find in this text a unified and detailed base that includes both analysis and synthesis techniques that can be easily expanded on by using a computer or programmable calculator. This book should be extremely useful as a college-level active-filter course book or supplemental text.

Chapter 1 begins with some basics, introducing such terms as damping, order, cascadeability, and other important filter concepts.

Chapter 2 tells all about operational amplifiers, particularly the circuits needed for active filters, and has a mini-catalog of suitable commercial devices.

The basic properties of the five elemental first- and second-order building blocks appear in Chapter 3.

Complete response curves that help us decide how much of a filter is needed for a certain job are the subjects of Chapters 4 and 5. The high-pass and low-pass curves of Chapter 4 cover seven different shape options through six orders of filter. The shape options represent a continuum from a Bessel, or best-time-delay, filter through a flattest-amplitude, or Butterworth, filter through various Chebyshev responses of slight, 1-, 2-, and 3-dB dips. The bandpass curves of Chapter 5 show us exactly what the response shape will be, again through the sixth order with five shape options. This chapter introduces a simple technique called *cascaded-pole synthesis* that greatly simplifies the design of active bandpass filters and gives you absolutely complete and well-defined response curves.

Actual filter circuits appear in Chapters 6 through 8. Four different styles of low-pass, bandpass, and high-pass circuits are shown. Low-pass and high-pass circuits include the simple and easily tuneable Sallen-Key styles, along with the multiple IC state-variable circuits. Bandpass versions include a single op-amp multiple-feedback circuit for moderate Qs and state-variable and biquad circuits for Qs as high as several hundred.

Chapter 9 shows us how to perform switching, tuning, and voltage control of active filters. It also looks at some fancier filter concepts such as allpass networks and bandstop filters and finally ends up with a very high performance ultimate-response filter called a *Cauer*, or *elliptic*, filter. Design curves through the fourth order are given.

Finally, Chapter 10 shows where and how to use active filters and gives such supplemental data as touch-tone frequencies, musical scale values, modem values, and so on, along with photos of here-and-now applications of the text techniques.

DON LANCASTER

This book is dedicated to the Bee Horse.

Contents

CHAPTER 1

Some Basics

A *filter* is a frequency-selective network that favors certain frequencies of input signals at the expense of others. Three very common types of filter are the *low-pass* filter, the *bandpass* filter, and the *high-pass* filter, although there are many more possibilities.

A low-pass filter allows signals *up to* a certain maximum frequency to be passed on; frequencies above this *cutoff* frequency are rejected to a greater or lesser degree. Hi-fi treble controls and turntable scratch filters are typical low-pass filters.

A bandpass filter selects a range of median frequencies while attenuating or rejecting other frequencies *above* and *below* those desired. The tuning dial on an a-m radio is an example of a *variable* bandpass filter.

Similarly, a high-pass filter blocks frequencies *below* its cutoff frequency while favoring those above. Hi-fi bass controls and turntable rumble filters are typical examples. Fig. 1-1 shows these filter characteristics.

Filters are an extremely important electronic concept. They are absolutely essential for radio, television, voice, and data communications. Telephone networks could not possibly exist without them. Audio and hi-fi systems need them, and electronic music finds them mandatory. Research tasks as diverse as seismology, brainwave research, telemetry, biomedical electronics, geophysics, speech therapy, new art forms, and process instrumentation all rely heavily on filter concepts.

Capacitors and inductors are inherently frequency-dependent devices. Capacitors more easily pass high frequencies and inductors better handle lower frequencies. Thus, most filters traditionally have been designed around combinations of inductors and capacitors. These are called *passive* filters.

Today, there is a new and often much better way to do filtering. Integrated circuitry, particularly the IC op amp, can be combined with resistors and capacitors to accurately simulate the performance of traditional inductance-capacitance filters. Since this new approach usually has gain and needs some supply power, filters built this way are called *active filters*. While active filters as a concept have been around for quite some time, only recently have reliable, easy-to-use circuits and simple design processes emerged.

This book is about these active filters and the design techniques behind them. It will show you the integrated circuits that can be used in active filters, the basic concepts of filter response shapes and how to get them, the circuits you can use to actually build the filters, the refinements you can add (such as tuning and voltage-controlled tracking), and finally, the application areas where active filters are now widely used.

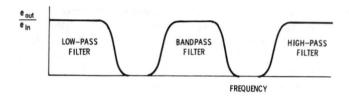

Fig. 1-1. Common types of filters.

WHY USE ACTIVE FILTERS?

There are many advantages of active filters, compared with traditional passive filters:

Low Cost—Component costs of active filters are usually far lower, particularly at very low frequencies, where inductors are large and expensive.

Isolation—Most active filters have very high input impedances and very low output impedances. This makes their response essentially independent of source and load impedances and their changes.

Cascadeability—Owing to the good isolation of active filters, complex filter problems are easily broken down into simple sections that combine to produce the desired final result.

Gain—Active filters can provide gain or loss as needed to suit system or filter requirements. Current gain is almost always provided; voltage gain is an option.

Tuning—Many active filters can be easily tuned over a wide range without changing their response shape. Tuning can be done electronically, manually, or by voltage control. Tuning ranges can go beyond 1000:1, much higher than is usually possible with passive circuits.

Small Size and Weight—This is particularly true at low frequencies, where inductors are bulky and heavy.

No Field Sensitivity—Shielding and coupling problems are essentially nonexistent in active filters.

Ease of Design—Compared with traditional methods, the methods explained in this book make the design of active filters trivially easy.

We might also consider what is wrong with active filters and what their limitations are:

Supply Power—Some supply power is needed by all active filter circuits.

Signal Limits—The operational amplifier used sets definite signal limits, based on its input noise, its dynamic range, its high-frequency response, and its ability to handle large signals.

Sensitivity—Variations in response shape and size are possible when component or operational-amplifier tolerances shift or track with temperature.

FREQUENCY RANGE AND Q

The range of useful frequencies for active filters is far wider than for any other filter technique. A frequency range of at least *eight decades* is practical today.

A useful lower frequency limit is somewhere between .01 and 0.1 Hz. Here capacitor sizes tend to get out of hand even with very high impedance active circuits, and digital, real-time computer filter techniques become competitive.

An upper limit is set by the quality of the operational amplifier used; something in the 100-kHz to 1-MHz range is a reasonable limit. Above this frequency, conventional inductor-capacitor filters drop enough in size and cost that they are very practical; at the same time, premium op amps and special circuits become necessary as frequency is raised. Still, the theoretical frequency range of active filters is much higher and even microwave active filters have been built.

If we consider using very simple circuits and very low cost operational amplifiers, active filters are pretty much limited to subaudio, audio, and low ultrasonic frequency areas. When bandpass filters are used, there is also a range for the narrowness of response that can be obtained. The inverse of the bandwidth of a single filter structure is called its Q and is simply the ratio of its bandwidth to its center frequency. A filter whose center frequency is 200 Hz and whose bandwidth is 2 Hz has a Q of 100.

With active filters, Q values of 500 or less are realizable. These higher Q values require active filter circuits with three or four operational amplifiers if the response is to be predictable and stable. Single operational-amplifier circuits are much more restricted in their maximum Q. A Q of 25 is an outside practical limit for these one-amplifier circuits. With today's dual and quad op-amp packages, using more than one op amp is not much more expensive than using a single one, particularly if you are guaranteed better circuit performance. Complete design details of both types of bandpass circuit will appear in Chapter 7.

A SIMPLE ACTIVE-FILTER CIRCUIT

Let us look at a simple active-filter circuit and see how it compares to its passive counterpart. Fig. 1-2 shows details. Active-filter circuits are rarely expected to "replace" one or more inductors on a one-for-one basis; instead, the overall mathematical response or circuit *transfer function* is considered and the filter is designed to *simulate* or *synthesize* this overall response. Instead of saying "this op amp replaces this inductor," we say "this active-filter circuit performs identically to (or better than) this passive inductor-capacitor filter." As long as the math turns out the same, we can get the same circuit performance, even if we cannot show exactly where the inductor "went."

Fig. 1-2A shows a low-pass filter made up of a series inductor, a shunt capacitor, and a load resistor. The reactance of an inductor increases with frequency, which makes it harder for high-frequency signals to reach the output. The reactance of a capacitor decreases with frequency, which means it has little effect on low-frequency signals but progressively shorts out or shunts to ground higher-frequency inputs. This *two-step* elimination process makes possible a *second-order* low-pass filter with a response that falls off as the *square* of frequency for high frequencies.

There are many possible response curves for this circuit. At very low frequencies, the gain of the circuit is nearly unity, since the reactance of the inductor is very low and that of the capacitor is very high. At very high frequencies, the two-step elimination process causes the response to fall off as the square of frequency, equal to a rate of 12 dB per octave.

We can call the point where the response starts to fall off significantly the *cutoff frequency* of the filter. The cutoff frequency is determined by the *product* of the inductor and capacitor values.

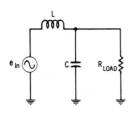 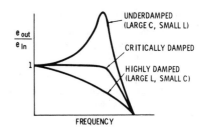

(A) Passive LC, second-order, low-pass filter and response.

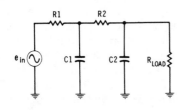 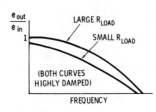

(B) Passive RC, second-order, low-pass filter and response.

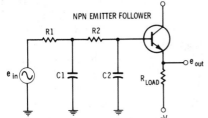 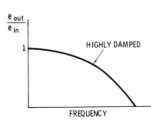

(C) Passive RC filter, emitter follower added to isolate load.

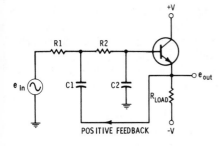 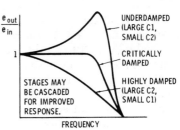

(D) Active RC filter and response identical to (A).

Fig. 1-2. Building a simple active filter.

11

We can control not only the product of the inductor and capacitor, but also their *ratio*. For instance, if the capacitor is very large and the inductor very small, the load resistor will not load the LC circuit much. The circuit behaves as a series-resonant circuit with relatively low losses, and so it will be on the verge of oscillation. For some frequencies near resonance, the circuit will exhibit voltage gain or *peaking*. This gives the *underdamped* response curve of Fig. 1-2A.

A more balanced ratio of load resistance, inductance and capacitance leads to a flatter response with no peaking or gain. The flattest of these is called a *critically damped* curve. If we go further and use a very small capacitor and a very large inductor, the load resistor dominates and gives a very droopy, *highly damped* response.

Note that these three responses all start out near unity gain and and end up dropping as the square of the frequency. Setting the damping by changing the inductor-capacitor *ratio* determines only the shape of the response curve near the cutoff frequency. The numerical value of the cutoff frequency is set by the *product* of the inductor and capacitor; the *damping* and the performance near the cutoff frequency are set by the *ratio* of the two.

We can add more capacitors and more inductors, picking up new products and ratios of component values. This lets us improve the response by increasing the *order* of the filter. Regardless of the order desired, the trick is to find a way to do this without inductors —and still get the same overall response.

Fig. 1-2B shows an approach using nothing but resistors and capacitors. Here there are two cascaded RC low-pass sections. Since two capacitors are shunting higher frequencies to ground, you would expect the *ultimate* falloff to also increase with the square of frequency. For high values of load resistance, you would also expect something near unity gain, perhaps a little less. So, this must be a second-order section with at least some responses similar to that of the inductor-capacitor circuit of Fig. 1-2A.

The problem is that the damping of this circuit is very high. The network is obviously lossy because of the two resistors. The damping is so bad that, instead of a reasonable shape near cutoff, there is a gradual falling off beyond the cutoff frequency. Still, this is a second-order section. Is there some way to repair the high damping by adding energy from a power source? Fig. 1-2C shows a first attempt in that direction.

In Fig. 1-2C, an emitter follower has been added to the output. This has eliminated any output loading effects since the emitter follower has unity gain, a high input impedance, and a low output impedance. Now, at least, the gain and damping are independent of the output load, even if the damping remains pitifully bad.

The key to building an active filter appears in Fig. 1-2D. Here the first capacitor is removed from ground and connected to the *output* of the emitter follower, so that there is *positive feedback* from the output back into the middle of the RC filter. This positive feedback bolsters the response and lets us reduce the damping to a point where we can get any acceptable response shape we like, just as we did by controlling the inductor-capacitor ratio of the passive filter of Fig. 1-2A.

What we have done is used energy from the power supply to make up for the losses in the filter resistors.

The positive-feedback connection delivers this excess energy back into the filter *only near the cutoff frequency*. This "localized" feedback is caused by the reactance of the feedback capacitor being too high to do much good at very low frequencies, and by the output signal being too small to be worth feeding back at very high frequencies. Thus, the response curve is boosted only near the cutoff frequency—the feedback does what it is supposed to do only where it is needed.

By changing the *ratio* of these two capacitors, the damping is changed, just as the response shape of the passive filter is controlled by using the ratio of inductor to capacitor. By the same token, the resistor-capacitor *product* sets the cutoff frequency, just as the inductor-capacitor product sets the passive frequency. The mathematical response of one circuit turns out to be *identical* to the response of the other circuit.

Note that one inductor and one capacitor set the frequency of the passive filter, while two capacitors and two resistors are needed to do the same job with the active filter. From an energy-storage standpoint, both circuits have two, and only two, energy-storage components—this is what makes them a second-order circuit.

What we have done is built an active filter that does exactly the same thing as a passive one, with the additional benefit of load isolation. At the same time, we have eliminated the cost, size, weight, and hum susceptibility of the inductor.

TYPES OF ACTIVE-FILTER CIRCUITS

Normally, we replace the simple emitter follower with an operational amplifier to pick up some other performance benefits. As many second-order sections as needed can be cascaded to get a desired overall response. Instead of cascading identical sections, the frequency and damping of each section is chosen to be a *factor* of the overall response. You can also tack on a first-order section (active or passive) consisting of a single resistor-capacitor pair to pick up odd-order filters such as a third- or fifth-order filter.

Figs. 1-3 through 1-6 show the basic second-order active-filter-section circuits that will be used in this text. When these are combined with other first- and second-order sections, following the guidelines of Chapters 4 and 5, virtually any desired overall filter response can be obtained.

Fig. 1-3 shows two low-pass circuits. The first of these is simply the circuit of Fig. 1-2D converted to an operational amplifier. It is called a *unity-gain Sallen-Key* circuit and it works on the principle of bolstering the response of two cascaded resistor-capacitor sections. Complete design and tuning details on all these sections will

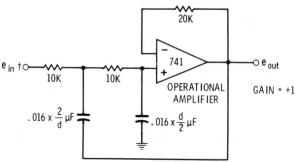

† must return to ground via low-impedance dc path.

(A) Unity-gain Sallen-Key.

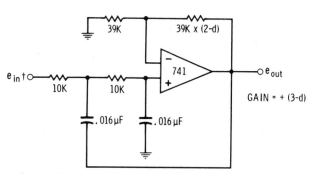

† must return to ground via low-impedance dc path.

(B) Equal-component-value Sallen-Key.

Fig. 1-3. Active low-pass filters, 1-kHz cutoff frequency, second-order.

be found later in the text; our interest here is simply to catalog what we are going to work with. The component values are shown for a 1-kHz cutoff frequency. In later chapters you will find that they are easily altered for any cutoff frequency you wish.

When the Sallen-Key math is examined in detail, it is seen that there is a "magic" gain value that makes everything very simple and independent of everything else. This magic gain value is $3 - d$, and d here stands for the damping we are after. The result is the *equal-component-value Sallen-Key filter* of Fig. 1-3B. This circuit features identical capacitors, identical resistors, easy tuning, and a damping set independently by the amplifier gain.

The bandpass circuits appear in Fig. 1-4. The Sallen-Key techniques do not really stand out as good bandpass circuits, so a slightly different circuit called a *multiple-feedback bandpass filter* appears in Fig. 1-4A. This circuit also uses an operational amplifier to bolster the response of a two-resistor, two-capacitor network, but does things in a slightly different way. The Q of the circuit is limited to 25 or less, and it turns out that any single-amplifier bandpass filter is limited to lower Q values.

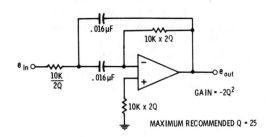

(A) Single-amplifier, multiple feedback.

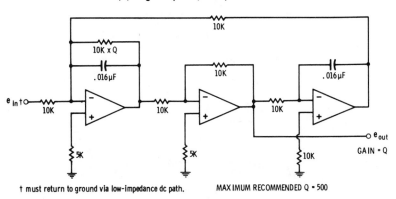

(B) Three-amplifier biquadratic.

Fig. 1-4. Second-order, active bandpass filters, 1-kHz resonance.

Fig. 1-4B shows a very interesting bandpass filter called a *biquad*. It supplies high-Q, single-resistor tuning if required, and an independent adjustment of frequency and *bandwidth* (not Q, but bandwidth). With this filter, Q values of 500 or higher are easily obtained, and the high performance and ease of design more than make up for the extra operational amplifier or two.

Fig. 1-5 shows the high-pass circuits. These are simply inside-out low-pass circuits that use a concept called *mathematical transformation by 1/f* to get responses that are mirror images in frequency to their low-pass counterparts or *prototypes*. The *single-amplifier Sallen-Key* filter is once again the simplest, while the *equal-component-value* filter offers easier design, circuit gain, and independent adjustment of practically everything. This filter is also switchable from high pass to low pass by interchanging components.

Finally, we consider a universal filter called a *state-variable filter*. It works on a different principle from the Sallen-Key types in that it uses two *integrators*, analog-computer style, to set up the analog or model of a pendulum. By picking the correct output, you can get a low-pass, a bandpass, or a high-pass response. In the bandpass mode, you get high Q, easy tuning, and a very simple design. Another benefit of this circuit is that it is easy to use electronic tuning or voltage control of the frequency over a wide range. Finally, the three outputs of the state-variable filter can be summed up to get all-pass, equalizer, bandstop, and other filter responses. Fig. 1-6 shows two state-variable filter versions.

More details of these circuits appear in Chapter 6 (low-pass), Chapter 7 (bandpass), and Chapter 8 (high-pass). In addition, complete, ready-to-go catalogs of "instant design" filters appear in

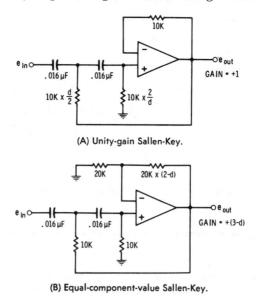

(A) Unity-gain Sallen-Key.

(B) Equal-component-value Sallen-Key.

Fig. 1-5. Active second-order high-pass filters, 1-kHz cutoff frequency.

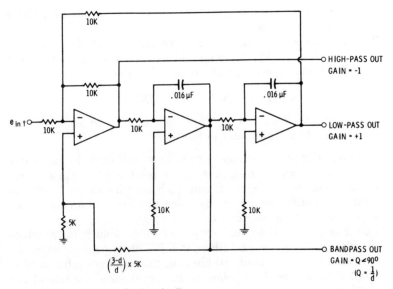

† must return to ground via low-impedance dc path.

(A) Unity-gain, state-variable filter.

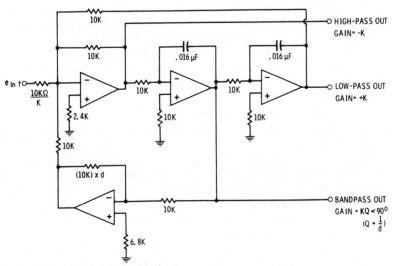

† must return to ground via low-impedance dc path.

(B) Variable-gain, state-variable filter.

Fig. 1-6. Active second-order universal filters, 1 kHz cutoff frequency.

Chapters 6 and 8, while the more advanced Cauer or elliptical filters are introduced in Chapter 9.

The rest of the book shows how to design and use these circuits, and how to decide what damping and frequency values are needed to get a desired response at a given cutoff frequency.

SOME TERMS AND CONCEPTS

Let us quickly review some important active-filter concepts that will be needed and used later in the book:

Cascading—High-performance active filters are built by cascading individual, noninteracting, first- and second-order sections. These sections are never identical, and each contributes its individual relative cutoff frequency and damping as a factor of an overall desired transfer-function response.

Cutoff Frequency—The cutoff frequency is the final point at which the filter response drops 3 dB or to 0.707 of its peak value on the way out of the passband. **All filters in this book are referenced to a cutoff frequency 3 dB below peak, regardless of the size of any passband lumps or the amount of filter delay.**

Damping—The damping of a second-order filter section is an index of its tendency toward oscillation. Practical damping values range from 2 to 0, with zero damping being the value for an oscillator, a damping of 1.414 being a critical value that gives maximum flatness without overshoot, and a damping of 2 being what you get when you cascade but isolate two resistor-capacitor networks that are identical. Highly damped filter sections combine to produce smooth filters with good overshoot and transient response. Slightly damped filters combine to produce lumpy filters with sharp rejection characteristics.

Decade—A 10:1 frequency interval.

Decibels—Decibels are a logarithmic way of measuring gain or loss. Decibels are defined as $20 \log_{10}$ of a voltage ratio. In active-filter work, decibels refer ONLY to a voltage ratio and *are totally independent of any impedance or standard reference level considerations.* A decibel chart appears as Fig. 1-7. When several stages are cascaded, their gain values multiply, but their decibel values simply add. Some useful relationships appear in Fig. 1-8.

Normalization—A normalized filter is one whose component values are adjusted to a convenient frequency and impedance level. A filter is easy to analyze if it is normalized to a frequency of 1 radian per second and an impedance level of 1 ohm. Designing with a filter is easy when the filter is normalized to a 10K-ohm impedance level and a 1-kHz cutoff frequency.

Order—The order of a filter governs the strength of its falloff with

dB	Gain	Loss	dB	Gain	Loss	dB	Gain	Loss	dB	Gain	Loss	dB	Gain	Loss
0	1.000	1.0000	4.0	1.585	.6310	8.0	2.512	.3981	12.0	3.981	.2512	16.0	6.310	.1585
.1	1.012	.9886	4.1	1.603	.6237	8.1	2.541	.3936	12.1	4.027	.2483	16.1	6.383	.1567
.2	1.023	.9772	4.2	1.622	.6166	8.2	2.570	.3890	12.2	4.074	.2455	16.2	6.457	.1549
.3	1.035	.9661	4.3	1.641	.6095	8.3	2.600	.3846	12.3	4.121	.2427	16.3	6.531	.1531
.4	1.047	.9550	4.4	1.660	.6026	8.4	2.630	.3802	12.4	4.169	.2399	16.4	6.607	.1514
.5	1.059	.9441	4.5	1.679	.5957	8.5	2.661	.3758	12.5	4.217	.2371	16.5	6.683	.1496
.6	1.072	.9333	4.6	1.698	.5888	8.6	2.692	.3715	12.6	4.266	.2344	16.6	6.761	.1479
.7	1.084	.9226	4.7	1.718	.5821	8.7	2.723	.3673	12.7	4.315	.2317	16.7	6.839	.1462
.8	1.096	.9120	4.8	1.738	.5754	8.8	2.754	.3631	12.8	4.365	.2291	16.8	6.918	.1445
.9	1.109	.9016	4.9	1.758	.5689	8.9	2.786	.3589	12.9	4.416	.2265	16.9	6.998	.1429
1.0	1.122	.8913	5.0	1.778	.5623	9.0	2.818	.3548	13.0	4.467	.2239	17.0	7.079	.1413
1.1	1.135	.8810	5.1	1.799	.5559	9.1	2.851	.3508	13.1	4.519	.2213	17.1	7.161	.1396
1.2	1.148	.8710	5.2	1.820	.5495	9.2	2.884	.3467	13.2	4.571	.2188	17.2	7.244	.1380
1.3	1.161	.8610	5.3	1.841	.5433	9.3	2.917	.3428	13.3	4.624	.2163	17.3	7.328	.1365
1.4	1.175	.8511	5.4	1.862	.5370	9.4	2.951	.3388	13.4	4.677	.2138	17.4	7.413	.1349
1.5	1.189	.8414	5.5	1.884	.5309	9.5	2.985	.3350	13.5	4.732	.2113	17.5	7.499	.1334
1.6	1.202	.8318	5.6	1.905	.5248	9.6	3.020	.3311	13.6	4.786	.2089	17.6	7.586	.1318
1.7	1.216	.8222	5.7	1.928	.5188	9.7	3.055	.3273	13.7	4.842	.2065	17.7	7.674	.1303
1.8	1.230	.8128	5.8	1.950	.5129	9.8	3.090	.3236	13.8	4.898	.2042	17.8	7.762	.1288
1.9	1.245	.8035	5.9	1.972	.5070	9.9	3.126	.3199	13.9	4.955	.2018	17.9	7.852	.1274
2.0	1.259	.7943	6.0	1.995	.5012	10.0	3.162	.3162	14.0	5.012	.1995	18.0	7.943	.1259
2.1	1.274	.7852	6.1	2.018	.4955	10.1	3.199	.3126	14.1	5.070	.1972	18.1	8.035	.1245
2.2	1.288	.7762	6.2	2.042	.4898	10.2	3.236	.3090	14.2	5.129	.1950	18.2	8.128	.1230
2.3	1.303	.7674	6.3	2.065	.4842	10.3	3.273	.3055	14.3	5.188	.1928	18.3	8.222	.1216
2.4	1.318	.7586	6.4	2.089	.4786	10.4	3.311	.3020	14.4	5.248	.1905	18.4	8.318	.1202
2.5	1.334	.7499	6.5	2.113	.4732	10.5	3.350	.2985	14.5	5.309	.1884	18.5	8.414	.1189
2.6	1.349	.7413	6.6	2.138	.4677	10.6	3.388	.2951	14.6	5.370	.1862	18.6	8.511	.1175
2.7	1.365	.7328	6.7	2.163	.4624	10.7	3.428	.2917	14.7	5.433	.1841	18.7	8.610	.1161
2.8	1.380	.7244	6.8	2.188	.4571	10.8	3.467	.2884	14.8	5.495	.1820	18.8	8.710	.1148
2.9	1.396	.7161	6.9	2.213	.4519	10.9	3.508	.2851	14.9	5.559	.1799	18.9	8.811	.1135
3.0	1.413	.7079	7.0	2.239	.4467	11.0	3.548	.2818	15.0	5.623	.1778	19.0	8.913	.1122
3.1	1.429	.6998	7.1	2.265	.4416	11.1	3.589	.2786	15.1	5.689	.1758	19.1	9.016	.1109
3.2	1.445	.6918	7.2	2.291	.4365	11.2	3.631	.2754	15.2	5.754	.1738	19.2	9.120	.1096
3.3	1.462	.6839	7.3	2.317	.4315	11.3	3.673	.2723	15.3	5.821	.1718	19.3	9.226	.1084
3.4	1.479	.6761	7.4	2.344	.4266	11.4	3.715	.2692	15.4	5.888	.1698	19.4	9.333	.1072
3.5	1.496	.6683	7.5	2.371	.4217	11.5	3.758	.2661	15.5	5.957	.1679	19.5	9.441	.1059
3.6	1.514	.6607	7.6	2.399	.4169	11.6	3.802	.2630	15.6	6.026	.1660	19.6	9.550	.1047
3.7	1.531	.6531	7.7	2.427	.4121	11.7	3.846	.2600	15.7	6.095	.1641	19.7	9.661	.1035
3.8	1.549	.6457	7.8	2.455	.4074	11.8	3.890	.2570	15.8	6.166	.1622	19.8	9.772	.1023
3.9	1.567	.6383	7.9	2.483	.4027	11.9	3.936	.2541	15.9	6.237	.1603	19.9	9.886	.1012

dB	Current or Voltage Ratio		dB	Current or Voltage Ratio	
	Gain	Loss		Gain	Loss
20.0	10.00	0.1000	85.0	1.778×10^4	5.623×10^{-5}
25.0	17.78	0.0562	90.0	3.162×10^4	3.162×10^{-6}
30.0	31.62	0.0316	95.0	5.632×10^4	1.78×10^{-6}
35.0	56.23	0.0178	100.0	10^5	10^{-5}
40.0	100.00	0.0100	110.0	3.162×10^5	3.162×10^{-6}
45.0	177.8	0.0056	120.0	10^6	10^{-6}
50.0	316.2	0.0032	130.0	3.162×10^6	3.162×10^{-7}
55.0	562.3	0.0018	140.0	10^7	10^{-7}
60.0	10^3	10^{-3}	150.0	3.162×10^7	3.162×10^{-8}
65.0	1.778×10^3	5.623×10^{-4}	160.0	10^8	10^{-8}
70.0	3.162×10^3	3.162×10^{-4}	170.0	3.162×10^8	3.162×10^{-9}
75.0	5.623×10^3	1.78×10^{-4}	180.0	10^9	10^{-9}
80.0	10^4	10^{-4}			

Fig. 1-7. Voltage decibel chart.

$$\text{Voltage decibels} = 20 \log_{10} \frac{e_{out}}{e_{in}}$$

- - - - - - - - - -

A ONE-decibel change is approximately a 10% change
A TWO-decibel change is approximately a 20% change
A THREE-decibel change drops to 70% amplitude
A SIX-decibel change is a 2:1 ratio
A TEN-decibel change is approximately a 3:1 ratio
A TWELVE-decibel change is a 4:1 ratio
AN EIGHTEEN-decibel change is an 8:1 ratio

- - - - - - - - - -

TWENTY decibels is a 10:1 ratio
THIRTY decibels is approximately a 30:1 ratio
FORTY decibels is a 100:1 ratio
FIFTY decibels is approximately a 300:1 ratio
SIXTY decibels is a 1000:1 ratio
EIGHTY decibels is a 10,000:1 ratio

- - - - - - - - - -

Ratios MULTIPLY; Decibels ADD

- - - - - - - - - -

Active-filter decibels are VOLTAGE RATIOS ONLY; they ignore impedance or reference levels.

Fig. 1-8. Some useful decibel relationships.

frequency. For instance, a third-order low-pass filter falls off as the *cube* of frequency for high frequencies, or at an 18-dB-per-octave rate. The number of energy-storage capacitors in most active filters determines their order. A fifth-order filter usually takes five capacitors, and so on. The higher the order of the filter, the better its performance, the more parts it will take, and the more critical the restrictions on component and amplifier variations.

Q—Q is simply the inverse of the damping and is used to measure the bandwidth of a second-order bandpass section. Practical Q values range from less than one to several hundred.

Scaling—Scaling is denormalizing a filter by changing its frequency or impedance level. Impedance level is increased by multiplying all resistors and dividing all capacitors by the desired factor. Frequency is shifted inversely by multiplying all frequency-determining resistors *or* by multiplying all frequency-determining capacitors by the desired factor. To double the frequency, cut all capacitor values by 2 *or* cut all resistor values by 2; do both and the frequency will be quadrupled.

Sensitivity—The sensitivity of an active filter is a measure of how accurate the component and operational-amplifier tolerances have to be to get a response within certain limits of what is desired. The sensitivity of the active filters in this book is normally quite good. Sensitivity guidelines appear in Chapters 4 and 5.

Shape Option—For a given-order filter, a wide variety of choices called shape options exist that determine the lumpiness of the passband, how fast the initial falloff will be, how bad the transient response will be, and so on. For low-pass and high-pass responses, this book gives you a choice of seven shape options called *best delay, compromise, flattest amplitude, slight dips, 1-dB dips, 2-dB dips, and 3-dB dips.* Shape options for bandpass filters include *maximum peakedness, maximum flatness, 1-dB dips, 2-dB dips, and 3-dB dips. Cauer* or *elliptical*-shape options appear in Chapter 9.

Transfer Function—The transfer function of an active filter is simply what you get out of the filter compared to what you put in. It is usually expressed as the ratio e_{out}/e_{in}. The transfer function usually includes both amplitude and phase information and sometimes is expressed in terms of a complex variable, "S." In this text, we will be concerned mostly with the amplitude response, and our main involvement with "S" will be as a notational convenience.

A DESIGN PLAN

Active-filter design is a multistep process, but one that can be made very easy if it is done one step at a time. First, we need to know more about operational amplifiers, particularly their theoretical circuit capabilities, along with the characteristics of commercial units. These are discussed in the next chapter.

Before an attempt is made to combine first- and second-order sections into composite responses, more details are needed on the circuits and mathematical properties of these basic building blocks. These are covered in Chapter 3. Chapters 4 and 5 show how to combine the basic building blocks into a filter of the desired response shape, with low-pass and high-pass filters appearing in Chapter 4 and bandpass filters in Chapter 5. Incidentally, the bandpass analysis techniques that are discussed are extremely simple and tell you the total response that you will get.

From that point, we get into the actual circuits, finding out what they do and how to get a desired response out of them. These are handled in Chapters 6 (low-pass), 7 (bandpass) and 8 (high-pass).

Chapters 6 and 8 also give you two catalogs of "ripoff" circuits which are immediately ready for your use without any mathematics.

Chapter 9 covers some practical points about component types, values, and tuning, along with two powerful techniques for voltage-controlled tuning. Finally, in Chapter 10, we will look at some of the products and concepts that use active filters, along with some specialized design data such as music scales, touch-tone frequencies, and so on.

CHAPTER 2

The Operational Amplifier

Integrated-circuit *operational amplifiers* are used as the basic gain block in most active filters. The op amp provides load isolation and a way to route energy from the supply in the proper amounts at the proper places in a resistor-capacitor network to simulate the energy storage of one or more inductors. The operational amplifier normally has a high input impedance and a low output impedance. This lets us cascade first- and second-order filter sections to build up a higher-order filter.

Op amps suitable for active-filter use are often based on the "741" style of device and its improved offspring. All of these circuits are very easy to use, and many cost less than a dollar.

Five ways are shown to use op amps in filters:

1. As a *voltage follower,* or as a unity-gain, high-input-impedance, low-output-impedance, noninverting amplifier.
2. As a noninverting *voltage amplifier,* providing a gain of 1 or more, with high input impedance and low output impedance.
3. As a *current-summing amplifier,* providing any desired gain, moderate input impedance, low output impedance, and signal inversion.
4. As a *summing block* that combines several input signals, some of which get inverted and some of which do not, a moderate input impedance, and a low output impedance.
5. As an *integrator* or "ramp generator" amplifier that mathematically gives us the integral, or the "area under the curve," of an input signal. Integrators usually invert the signal and have a moderate input impedance and a low output impedance.

A CLOSER LOOK

Fig. 2-1 shows the internal schematic of a typical 741 type of operational amplifier. The op amp is normally powered by a dual or split power supply ranging from ±5 to ±15 volts. There are *two* inputs, each usually consisting of the base of an npn transistor. The internal circuitry gives high amplification of the *difference* voltage between these two inputs.

The input having a phase the same as the output at low frequencies is called the *noninverting* input and is identified by a + symbol. The input having a phase the opposite or 180° from the output is called the *inverting* input and is shown as a − on the symbol. A positive input step on the noninverting, or +, input drives the output *positive;* a positive input step on the inverting, or −, input drives the output *negative*.

After input amplification, the response of the op amp is *compensated* or stabilized by an internal 30-pF capacitor. This compensation ensures amplifier stability for any reasonable set of circuit connections. After compensation, the amplified signals are routed to an *output* stage that provides a low output impedance and considerable drive power.

The typical gain of an op amp at low frequencies is over 100,000, but this drops rapidly as frequency is increased. Because of this

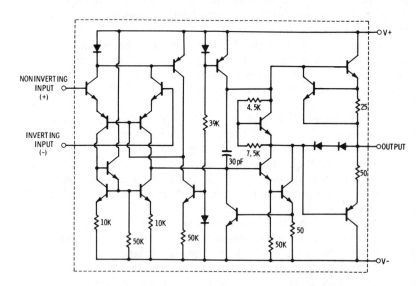

Fig. 2-1. A type 741 operational amplifier. Note that dc bias current must be provided for *both* inputs.

very high gain, an op amp is NEVER run "wide open." You always place resistive or capacitive *negative feedback* around the circuit from the output to the inverting (−) input.

If the feedback is properly placed, the circuit gain or its response function will be set ONLY by the input and feedback resistors, and the response will be essentially *independent* of the actual amplifier gain, the supply voltage, or temperature effects.

Note that the *inputs* to the operational amplifier are usually by way of the bases of two npn transistors. A small bias current must be provided, and any differences between these transistors or their external source resistors will produce a bias current difference called an *offset* current. It is ABSOLUTELY ESSENTIAL that there be a direct-current return path to ground or another stable dc bias point for BOTH inputs at all times. Ideally, these return paths should be of identical impedance value to minimize offsets caused by differential bias drops. Resistors in the 10K to 100K region are often a good choice for working values.

Fig. 2-2 sums up the operating rules for an operational amplifier. The op-amp circuits that follow simply will not work unless these three rules are obeyed.

An operational amplifier circuit will not work at all unless:

1. External feedback limits the gain or desired response to a design value.

2. Both inputs have a direct-current return path to ground or a similar reference.

3. The input frequencies and required gain are well within the performance limitations of the op amp used.

Fig. 2-2. Some important op-amp rules.

SOME OP-AMP CIRCUITS

The Voltage Follower

The *voltage follower* of Fig. 2-3 can be thought of as a super emitter follower. It has unity gain, a very high input impedance, and a very low output impedance, and it does not invert. The circuit has its inverting input connected to its own output. The noninverting input receives the signal. The high gain of the amplifier forces the difference between the two inputs continually to zero, and the output thus *follows* the input at identical amplitude.

The load presented by the + input on the active-filter circuitry is very light, yet the output of the amplifier can drive a substantial or changing output load without any changes getting back to the input and altering the response of the filter.

The feedback resistor from the output to the inverting input is not particularly critical in value. Usually it is chosen to provide an offset identical to that of the input by making its value identical to the impedance seen back to ground through the active-filter circuit. Optionally, this resistor can also be adjusted to provide a way of zeroing the small output offset voltage.

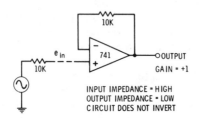

Fig. 2-3. The voltage follower.

Voltage followers are used to isolate a load or to obtain a high input impedance and a low output impedance. Note that a dc return path must be provided back to ground through an active filter circuit, even though only a light loading at the + input exists. Impedance levels in the 10K to 100K range are usually recommended. Values below these are hard to drive, and values above these tend to introduce offset and offset adjustment problems.

A single transistor connected as an emitter follower could also be used as a high-input-impedance, low-output-impedance, "unity-gain" amplifier. Its limitations include a gain always slightly less than one, a temperature-dependent, 0.6-volt, output offset voltage, and a lower ratio of input to output impedance. For most applications, the price difference between the two circuits is negligible and the op-amp voltage follower is the better choice.

Voltage Amplifier

Fig. 2-4 shows how to add one resistor to a voltage follower to get a noninverting amplifier with gain. Instead of the output going back directly to the inverting input, it goes back through a voltage divider. The *step-down* ratio of this divider determines the circuit gain. We retain a high input impedance and a low output impedance.

Suppose the feedback resistor, R, is 22K. What will the gain be? Since R is 22K, the total divider resistance will be 22K + 10K, or 32K. Only 10/32 of the output will get back to the inverting input to match the input signal, so the gain will turn out to be 32/10, or

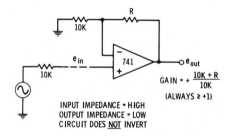

Fig. 2-4. The noninverting amplifier with gain.

3.2. Depending on R, we can get any positive gain we like from unity up to values near the open-loop gain of the op amp at the frequencies of interest. Note that there is no way to reduce the gain of this circuit *below* unity. As with the voltage follower, the + input only lightly loads the active-filter circuitry preceding it, but a dc return path to ground through the active filter is essential if the circuit is to work.

The values of the voltage-divider resistors are not particularly critical, although their *ratio* is, since the ratio sets the gain. Usually, you arrange things so that the *parallel combination* of the voltage-divider resistors is about the same as the impedance looking back on the noninverting input. This minimizes balance-and-offset problems.

Current-Summing Amplifier

A unity-gain current-summing amplifier appears in Fig. 2-5. While it looks about the same as the earlier circuits, its behavior is very much different. In this circuit, the operational amplifier is forced by feedback to present an extremely *low* apparent input impedance at the − or inverting input. Ideally, this very low impedance is ZERO. The concept is called a *virtual ground.* Let us see why it exists and why this circuit inverts the input signal.

The + input is essentially at ground, since the base current through the 5K resistor will give us a drop of only half a millivolt or

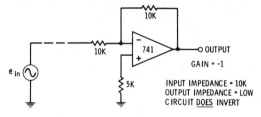

Fig. 2-5. Unity-gain inverting amplifier.

so. Suppose that a small positive-going input is applied to the left end of the 10K input resistor. This signal gets strongly amplified and drives the output *negative*. The negative-going output is fed back through the top or *feedback* resistor and *continuously attempts to drive the voltage on the − input to ground*. Looking at things a bit differently, only a negligible input current actually goes *inside* the op amp at the − input, so current through the input resistor must also go through the feedback resistor. If the voltage at the − input is not zero, the very high gain of the operational amplifier strongly amplifies the error difference. The error difference is fed back to drive the − input continuously to ground. Since the − input is always forced very near ground regardless of the size of the input signal, the −input is a virtual ground point.

The operational amplifier continuously forces its inverting − input to ground. Since the same current must flow through input and feedback resistors, and since both these resistors are the same size, the output will follow the input, but will be its *inverse*, swinging positive when the input goes to ground, and vice versa. If the input is a low-frequency sine wave, it will undergo a 180-degree phase reversal while going through the amplifier.

Since the − input is a virtual ground, the input impedance will be determined *only* by the input resistor. In this circuit, the input impedance is 10K. As with virtually all op-amp circuits, a dc bias return path MUST be provided back through the active-filter circuitry and must eventually reach ground.

Once again, the value of the resistor feeding the noninverting + input is not particularly critical. It is often chosen to be equal to the *parallel equivalent* of the *total* of the resistors on the − input, as this value minimizes bias current offset effects.

The 10K input resistor could theoretically be split up any way between the previous circuit source impedance and a fixed input resistor. For best overall performance, use a fixed input resistor and the lowest possible source impedance. In this way, source impedance variations and drifts are minimized.

A variable-gain inverting amplifier is shown in Fig. 2-6. The input current must equal the feedback at all times, minus a negligible input bias current. The − input essentially sits at a virtual ground. So, by making the feedback resistor larger or smaller, we can get any gain we want. If the feedback resistor is doubled, the output voltage swing must double to provide for the same input current as before; the gain then becomes −2 (the gain is negative because of circuit inversion). If the value of the feedback resistor is divided by 4, the gain drops to −1/4, and so on.

The gain obtained turns out to be the negative *ratio* of the *feedback* resistor to the *input* resistor. As long as this ratio is much less than the amplifier open-loop gain at the frequencies of interest, *the circuit gain will be completely determined by the resistor ratio and will be independent of temperature, op-amp gain, or supply voltage.*

If we like, we could hold the feedback resistor constant and vary the input resistor. With this design procedure, the gain varies inversely with input resistance. The input impedance also changes if

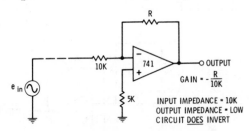

Fig. 2-6. Variable-gain inverting amplifier.

you change the input resistor. Note that gain values of greater or less than unity are obtained, depending only on the ratio of the two resistors.

A summing amplifier with two inputs is shown in Fig. 2-7. Since each input resistor goes to a virtual ground, there is no interaction between inputs, and the gain of each input is set by the ratio be-

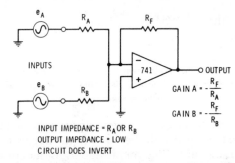

Fig. 2-7. Two-input, inverting, current-summing amplifier.

tween itself and the common feedback resistor. As usual, dc return paths must be provided back through the input circuitry. The value of the resistor on the noninverting input is not critical and often may be replaced with a short circuit. Its optimum value equals the *parallel combination* of *all* resistors on the − input. This optimum value gives minimum offset.

The input impedance of each input is simply the value of the input resistor, since all resistors on all inputs go to a virtual ground.

Summing Block

The inputs on the + and − pins of the op amp can be combined to get many simultaneous combinations of inverting and noninverting inputs and gains. One thing that cannot be done is to make the gains of all the inputs independently adjustable if there is only one op amp and there are mixed inverting and noninverting signals. One circuit is shown in Fig. 2-8.

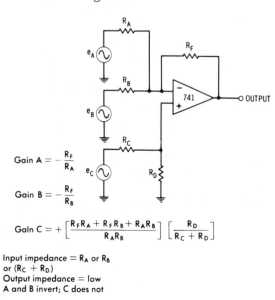

$$\text{Gain A} = -\frac{R_F}{R_A}$$

$$\text{Gain B} = -\frac{R_F}{R_B}$$

$$\text{Gain C} = +\left[\frac{R_F R_A + R_F R_B + R_A R_B}{R_A R_B}\right]\left[\frac{R_D}{R_C + R_D}\right]$$

Input impedance = R_A or R_B
or ($R_C + R_D$)
Output impedance = low
A and B invert; C does not

Fig. 2-8. A summing block that inverts two inputs but does not invert the third. Gains cannot be independently adjusted without interaction. Note that gain "C" is not an obvious expression.

You can analyze this circuit by assuming you have only one input at a time and have *shorted* all the other input signals to ground.

The gain of the inverting input signals is independent of the signal applied to the noninverting (+) input, as long as the inverting input somehow obtains its bias and as long as operation is within a frequency range where there is lots of extra open-loop gain. Thus, the *inverting*-input signals behave just as they did before; their gains are simply a ratio of their individual input resistances to the feedback resistance.

The gain applied to a signal by a *noninverting* input is by no means an obvious thing; in fact, it turns out to be a rather complex expression. There are two things that enter into the gain of non-inverting input C: the voltage attenuation of the divider R_C and R_D, and the actual gain of the circuit looking from the + input as determined by R_A, R_B, and R_F.

If input signals A and B are temporarily removed and replaced with short circuits, the positive voltage gain from the + input is set by a voltage divider. This voltage divider consists of the feedback resistor, R_F, and the *parallel* combination of R_A and R_B. The final gain expression appears in Fig. 2-8, and is quite complex. There is also a limit to the gain values you can have, as the gain from the + input to the output always has to be unity or greater.

This may seem like a strange circuit, but it is very useful for the *state-variable* input summing block. Its main limitation is that the gain of A or B cannot be changed without the C gain also changing in an obscure way. You *can* change the gain from input C without changing A or B, though, simply by varying resistor R_C. We will encounter this circuit in several places later in the text.

If we want three independently variable, noninteracting inputs, two of which invert and one of which does not, one simple approach is to add a second op amp to *invert* the signal we eventually want to end up with in noninverted form. This can be followed with a three-input version of Fig. 2-7, and all three gains can be simply and independently adjusted.

The Integrator

If the feedback resistor of a one-input inverting op-amp circuit is replaced with a capacitor (Fig. 2-9), an operational amplifier is

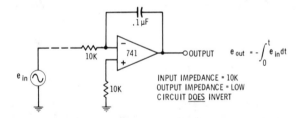

$$e_{out} = -\int_0^t e_{in} dt$$

INPUT IMPEDANCE = 10K
OUTPUT IMPEDANCE = LOW
CIRCUIT <u>DOES</u> INVERT

Fig. 2-9. Integrator. The rate of integration is set by input resistor and feedback capacitor.

converted into an *integrator* or an "area-under-the-curve" type of device. This happens because the capacitor has a memory or an ability to store the previous history of current/time variations, through the use of its stored charge.

Suppose that the charge on the capacitor is zero and that we apply a positive input current. The op amp will always adjust its output to provide an offsetting or cancelling input current, since any current that goes through the input resistor must come out of the capacitor if a virtual ground is to be held at the − input. If the left end of the resistor has a positive input voltage on it and if the right end is at a virtual ground, then the current through the resistor must be constant. The current through the capacitor must also be constant, so the capacitor will have to charge linearly in the *negative* direction. The output voltage will be a negative-going ramp. If we look at the input signal as a voltage-versus-time curve, the capacitor will be providing an output voltage that represents the time-voltage *integral* or the total of the history of the time-voltage variations.

Charging cannot go on forever, for eventually the op amp will *saturate* as it nears its negative supply. We have to pick a suitable *time constant* of the input resistor and feedback capacitor to match the input-signal frequencies; otherwise the op amp will eventually saturate on either the positive or the negative swing.

Fig. 2-10 shows two typical integrator waveforms. With a square-wave input, we get a negative triangle-wave output, since a triangle

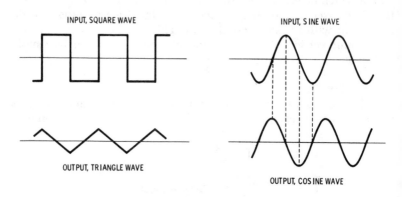

Fig. 2-10. Typical integrator waveforms.

wave represents the time average "area under the curve" of a square wave. We have to pick a suitable time constant to match the frequency of the input. If the time constant is too long, the output triangle wave will be very small; if the time constant is too short, the output triangle wave will try to get so big that the op amp will saturate near the supply limits.

If the input to the integrator is a sine wave, an interesting thing happens. A sine wave appears at the output, but its phase is shifted by 90 degrees, so it is really a *cosine* output. If the time constant exactly equals the radian input frequency, the output amplitude will equal the input amplitude but will be shifted in phase by 90 degrees. A 1-ohm resistor and a 1-farad capacitor would give a theoretical time constant of one second; a 10K resistor and a .016-

microfarad capacitor would give a time constant useful in a 1-kHz-cutoff active filter.

A simple sine-wave oscillator can be built out of two integrators and an inverter, as shown in Fig. 2-11. This is the electronic analog of a pendulum. Suppose that we start with a sine wave from some unspecified source that is the right frequency and amplitude. If it is integrated with the right time constant, a new sine wave is obtained that is actually a cosine wave that is shifted in phase by 90 degrees. We also get an inversion since the sine wave was applied to the − input. Suppose that we integrate again, still with the right

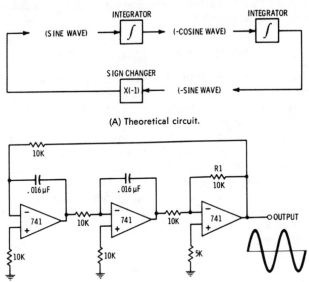

(A) Theoretical circuit.

(B) Practical 1-kHz oscillator. R1 may have to be adjusted to ensure starting and a stable level.

Fig. 2-11. Building a sine-wave oscillator pendulum analog with two integrators and an inverter.

time constant. This picks up another 90-degree phase shift, giving 180 degrees of phase shift and two inversions. The two inversions cancel out, so we end up with an inverted replica of the input. We route this input replica to a stage with a gain of −1, and end up with an output signal that looks just like the input. To build an oscillator, simply connect output to input. This gives the analog of a lossless pendulum.

This circuit is widely used to generate low-distortion sine waves, although some technique to stabilize amplitude must be used. A practical solution to the amplitude problem appears in Chapter 10.

Now, an oscillator is normally not a good filter, since we obviously do not want any output if no input has been applied. If some rust or some air resistance is added to the pendulum, we get the equivalent of a mechanical filter. To do this electronically, we must add some *damping* to the circuit. This can be done in two ways.

In the first way, damping is added by putting a resistor directly across one of the capacitors. This gives a circuit called a *biquad,* useful as an active bandpass filter or an electronic chime or ringing circuit. In the second way, feedback is electronically added from the other integrator. This, too, behaves as damping, leading to a *state-variable* filter that is useful as an active low-pass, bandpass, high-pass, or special-purpose filter.

SOME OP-AMP LIMITATIONS

Several restrictions must be observed if an op amp is to perform as expected. These restrictions include the frequency response of the op amp, its *slew rate,* its input noise, its *offset* voltages and currents, and its dynamic range.

At the highest frequency of operation, there must be available enough *excess* gain to let the feedback resistors behave properly. Excess gain also usually ensures that internal op-amp phase shifts will not introduce impossible problems.

Data sheets for any operational amplifier always show the frequency response. A quick look at a 741 data sheet (Fig. 2-13) shows that the frequency response is already 3 dB down from its dc value at 6 hertz! From this point, it continues to drop 6 dB per octave, ending with unity gain near 1 MHz. A reasonable guideline is to make sure that the op amp you are using has at least *ten times,* and preferably twenty times, the gain you are asking of it at the highest frequency of interest. In the case of a low-pass or a bandpass filter, you normally are not very interested in frequencies much above the passband. On the other hand, with a high-pass filter, the op amp must work through the entire passband of interest. The frequency response of the op amp will set the *upper* limit of the passband, while the active high-pass circuit will set the *lower* passband limit.

With an integrator, excess gain is still needed, but just how much extra depends on the circuit. A reasonable minimum limit is *three to five times* the Q expected of the circuit at the highest frequency of interest.

The maximum frequency of operation depends both on the op amp and the circuit. Specific maximum limits are spelled out later in the text.

The *slew rate* is a different sort of high-frequency limitation. It can be a limit much more severe than the simple frequency-response limitations. The slew rate places a limit on how fast the *large-signal* swing can change. Thus, the *large-signal* response is often much worse than the small-signal high-frequency response.

For instance, if the slew rate is 0.5 volt per microsecond, and we have a 3-volt peak-to-peak signal, it will take 6 microseconds to change 3 volts at 0.5 microsecond per volt. This will take care of one half of the signal swing, and the other half will go the other way. So, 12 microseconds is the highest frequency period we can handle at a 3-volt level, or at 80 kHz or so equivalent frequency. As the output amplitude goes *up*, the maximum possible operating frequency goes *down* because of this slew-rate limitation. The slew-rate limit is independent of the open-loop gain and is just as much or more of a problem at unity gain as at higher values.

Premium op amps have much higher slew rates than the common 741-type devices and are much more useful at higher frequencies. We will look at some actual figures in just a bit. For many upper-frequency audio applications, the conventional 741 has too low a slew rate to be genuinely useful.

Offset effects can involve both voltages and currents and can be both internal to and external to the op amp. An offset current or voltage gives a nonzero output voltage for a zero input voltage. Normally, the op amp itself has its offset matched to within a few millivolts, referenced to the input. Sometimes there are extra pins on an op-amp package that let you null the offset to a minimum value. Even when the offset is nulled, there is still some temperature dependence, so the best you can usually hope for is several hundred microvolts of offset.

You end up introducing additional offset whenever the input base-bias current provides a voltage drop across an input resistor. If the voltage drops on both the + and − inputs can be made identical, these offsets can be made to nearly cancel.

For instance, a typical 741 input-bias current is something under 0.1 microampere. This current gives a millivolt drop across a 10K resistor, a 10-millivolt drop across a 100K resistor, and 0.1-volt drop across a 1-megohm resistor. If you keep the resistors on the inputs nearly identical and below 100K, offsets should not be a serious problem. You will have to watch more closely at higher impedance levels, however. Note that offset is referred to the input. With a gain of 3, you get 3 times the output offset. With a gain of 100, you get 100 times the input offset at the output.

The operational-amplifier *noise* is the output signal you get in the absence of an input signal, again referenced to the input. A typical input broad-band noise value would be 10 microvolts; at a gain of 3, you would get 30-microvolts output, and so on. This noise sets a definite lower limit to the size of the input signals, depending on the signal-to-noise ratio you want. For many active-filter problems, input noise is not a serious restriction. Premium low-noise devices are available where ultrasmall input signals must be used.

The *dynamic range* of an op amp is the ratio of the largest useful signal to the smallest. Obviously, the system signal levels should be adjusted to make sure they are in the center of the dynamic range of the amplifier being used; otherwise you will be seriously restricted in terms of input signal amplitudes. The *minimum* signal you can handle is set by input noise or offset problems. About 5 millivolts of output at unity gain is a safe lower limit. The *maximum* signal you can handle is set by the supply voltages and by how near to the supply the op-amp output can come without distorting. With ±15-volt supplies, up to 5 volts rms can usually be handled. There is then a safe 1000:1 or 60-dB dynamic range at your disposal. Forty deci-

---

1. The OUTPUT SIGNAL LEVEL should lie between 15 millivolts rms and 1.5 volts rms.

2. The OPEN-LOOP GAIN should be at least ten times the desired circuit gain at the highest frequency of interest. On an active filter using integrators the open-loop gain should be a minimum of five times the circuit Q.

Circuit Gain	Minimum Op Amp Gain
1	10
3	30
$2Q^2$	$20Q^2$
Integrator	$5Q$

3. The op-amp SLEW RATE sets the following limits on the highest 1-volt rms sine-wave frequency:

Slew Rate	Max 1-Volt Sine-Wave Frequency
.5 V/μsec	80 kHz
1 V/μsec	160 kHz
70 V/μsec	10 MHz

4. Input BIAS CURRENTS generate around one *millivolt* per 10K impedance in a 741 type amplifier.

Fig. 2-12. Some guidelines for operational-amplifier circuits.

bels of dynamic range is more than enough for many routine applications; so if about 10 dB is lopped off from both ends, the *optimum* working signal level at the op-amp output should range from 15 millivolts to 1.5 volts rms.

Some rule-of-thumb use and design rules for operational amplifiers appear in Fig. 2-12.

WHICH OPERATIONAL AMPLIFIER?

Four very good operational amplifiers for active-filter use appear in Figs. 2-13 through 2-16. Each figure shows the pin connections and important parameters, such as the gain versus the frequency.

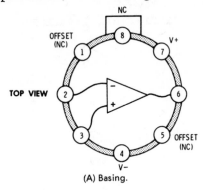

(A) Basing.

SUPPLY VOLTAGE RANGE ± 5 TO ± 18 V
SLEW RATE: 0.5 VOLT/ μSEC
NOISE: 10 μVOLTS, 100 kHz Δf

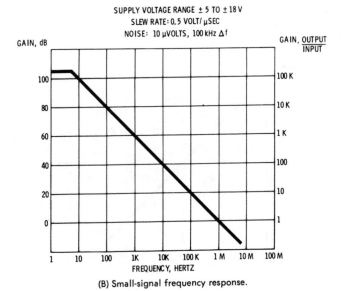

(B) Small-signal frequency response.

Fig. 2-13. Characteristics of the μA741 operational amplifier.

Fig. 2-13 shows the μA741 manufactured by *Fairchild* and others. This is the original easy-to-use op amp and is probably the most widely available one as well. Its internal compensation makes it very easy to use and very low in cost. Its typical price is less than a dollar, with surplus units as low as 35¢ each.

On the debit side, the 741 has a rather poor frequency response and an extremely bad slew-rate limit of half a volt per microsecond. The 741 is limited in usefulness above 10 kHz or in midaudio high-Q active filters.

The 741 has an internal input-offset adjustment using pins 1 and 5. These are usually left open in active-filter use.

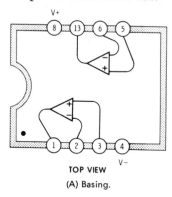

TOP VIEW

(A) Basing.

SUPPLY VOLTAGE RANGE ± 5 TO ± 18 V
SLEW RATE: 0.5 VOLT/ μSEC
NOISE: 10 μVOLTS, 100 kHz Δf

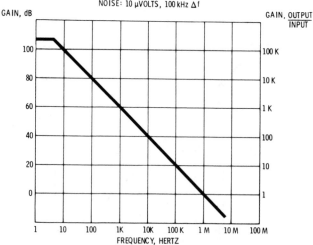

(B) Small-signal frequency response.

Fig. 2-14. Characteristics of the 5558 dual operational amplifier.

Fig. 2-14 shows the Signetics 5558, a dual-741 in an 8-pin minidip package. While its performance is essentially identical (no offset provisions) to the 741, it provides two amplifiers in an 8-pin minidip package at very low cost and circuit size.

You can also get quad 741s in a standard 16-pin DIP package. The Raytheon 4136 of Fig. 2-15 is typical and is particularly handy for state-variable and other multiple op-amp applications. A pair of 5558s take up the same amount of room as a single 4136. Again, there are no offset adjustments.

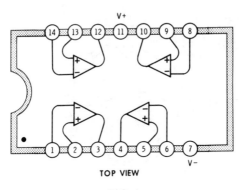

TOP VIEW

(A) Basing.

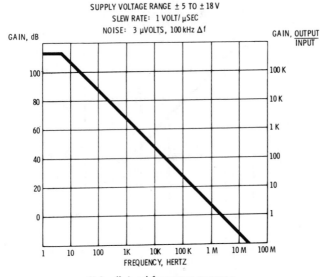

(B) Small-signal frequency response.

Fig. 2-15. Characteristics of the RM4136 operational amplifier.

There are several "second generation" 741s with improved slew rates and slightly better frequency response. The typical slew rate of these improved devices is around 5 volts per microsecond, enough better that these devices are good throughout the high-frequency audio range. Typical devices go by 741S or 741HS numbers and are manufactured by Motorola, Silicon General, and others. The Raytheon 4558 is an improved version of the 5558, with higher gain and slew rate and lower bias currents. If you use this device, note that the inputs are *pnp* transistors and the bias current sense is backward.

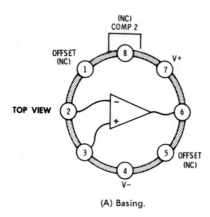

(A) Basing.

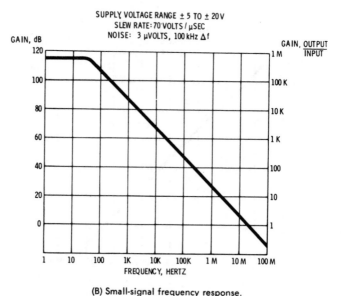

(B) Small-signal frequency response.

Fig. 2-16. Characteristics of the LM318 operational amplifier.

Designing op-amp active-filter circuits—some

A. For an equal-component-value Sallen-Key low-pass section, provide a noninverting, high-input-impendance, low-output-impedance amplifier with a gain of 3 — d where d is some number between zero and two.

We use the circuit of Fig. 2-4. At a 10K impedance level,

$$\text{Gain} = 3 - d = \frac{10K + R}{10K}$$

$$(3 - d) \ 10K = 10K + R$$

$$R = 10K \ (3 - d - 1)$$

$$R = 10K \ (2 - d)$$

The final circuit looks like this:

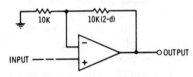

Thus a feedback resistor of normalized value 2 — d is needed in this circuit to get a gain of 3 — d overall.

B. For a state-variable universal filter, provide a summing block with two inverting unity-gain inputs and a noninverting input with a gain of "d," where d is some number between zero and two.

We use the circuit of Fig. 2-8. Pick $R_F = 10K$. R_A and R_B will also be 10K. The other two resistors are somewhat more difficult to calculate. Let R_D be 5K for optimum offset performance. We now have to calculate resistor R_C:

$$\text{Gain C} = \text{"d"} = \left[\frac{(10K)(10K) + (10K)(10K) + (10K)(10K)}{(10K)(10K)} \right] \left[\frac{5K}{R_C + 5K} \right]$$

$$d = [3] \left[\frac{5K}{R_C + 5K} \right]$$

$$dR_C + d5K = [3]5K$$

$$R_C = 5K \left[\frac{3 - d}{d} \right]$$

and the final circuit looks like this:

This way, a feedback resistor of normalized value (3 — d)/d is needed to get an overall gain of +d.

Fig. 2-17.

41

The National LM318 of Fig. 2-16 is a much more sophisticated amplifier than the 741, despite its identical pin connections. It has a slew rate of 70 volts per microsecond and a unity-gain frequency response of approximately 30 megahertz. It is priced around three to four times the cost of a 741, with a $4 unit price being typical. It is an excellent choice for all high-frequency active-filter applications, although the one-per-can limitation can be restrictive when many op amps are needed in a fancy filter circuit.

All of these amplifiers are *internally compensated*. This makes them very easy to use, but it severely limits the slew rate and frequency response. There are lots of uncompensated op amps available. These are usually much faster, but in exchange for this you have to add external parts yourself to stabilize the amplifier. One surprise as you start using these uncompensated amplifiers is that the *lower* the gain you are after, the *harder* the amplifier will be to stabilize. Suitable units are offered by Texas Instruments, Harris Semiconductor, Advanced Micro Devices, and others. The LM318 has optional external compensation—you can improve its response with external components if you need slightly better performance. The RCA 3130 is a low-cost, field effect, CMOS op amp with a better performance than a 741 and incredibly lower input currents. It is stabilized with a 50 pF capacitor. Much higher impedance levels may be used with a 3130, particularly at very low filter frequencies.

Another type of amplifier is the "automotive op amp," or Norton amplifier. It is low in cost and available in quads and often operates from a single supply. It is NOT a true traditional operational amplifier. This type is tricky to use and needs vastly different biasing than is shown in this chapter. It cannot be "one-for-one" dropped into the circuits of this text. Use this type only if you are thoroughly familiar with it.

Better op amps are coming along daily. But at this writing, the 4558 is the best choice for simple active filters in terms of size and economy; the LM318 should be used at higher frequencies or where high performance is needed; the 3130 is a good choice for low-frequency, high-impedance work.

Fig. 2-17 shows two examples of operational-amplifier circuit design.

CHAPTER 3

First- and Second-Order Networks

Complex active filters are normally built up by *cascading* two relatively simple types of circuits called *first-order* and *second-order* networks. If we choose the right combination of values for these networks, we get an overall response curve that does a more complex filter task. By using active techniques, we prevent the cascaded sections from interacting. Methods of this chapter and the following two can be used to select the right values for each part of each network to get the composite filter result that is desired.

There are only two first-order networks, a high-pass one and a low-pass one. All you can control on these is the center frequency and the impedance level. There are seven possible second-order networks. Of these, the three most popular are a low-pass, a bandpass, and a high-pass response. The others can be obtained by summing these three in different ways.

On a second-order section, it is possible to control the impedance level, the center frequency, and a new feature called the *damping*, d, or its inverse, Q. Damping or Q sets the peaking or droop of the response at median frequencies near the cutoff frequency.

A first-order section is not very useful by itself as a filter. Some second-order sections make good limited-performance filters. Better-working active filters use combinations of first- and second-order sections in cascade, perhaps combining two second-order sections for a fourth-order response, or two seconds and a first for a fifth-order filter, and so on.

In this chapter, we will first look at an important work-saving concept called *normalization* and its related technique, *scaling* or denormalization. Then, we will look at the basic properties of all the major first- and second-order sections. Chapters 4 and 5 will show how to combine these sections into useful filter response curves, while the following chapters show exactly how to build the circuits to do these tasks.

NORMALIZATION AND SCALING

Fortunately, it is necessary to analyze a certain filter only once. After that, a simple multiplication or division of component values can be used to *shift* the filter to any desired impedance level or center frequency. In this way, we do not have to start from scratch every time a new filter design is needed.

Since we have a choice of frequencies and impedance levels, it pays to choose the simplest possible one for *analysis.* Later on, we should choose the simplest possible one for *synthesis,* or actual usage. Both of these simplified circuits are *normalized* ones. Denormalization to get a final filter is called *scaling.*

Analysis is easiest to do on a circuit with a cutoff frequency of one radian per second and an impedance level of one ohm. The time constant of a 1-ohm resistor and a 1-farad capacitor has a frequency equivalence of one radian per second. Synthesis is easiest to do when the circuit is designed to a 10K impedance level and a 1-kHz cutoff frequency. The time constant of a .016-microfarad capacitor and a 10K resistor is equivalent to a 1-kHz cutoff frequency.

Fig. 3-1 shows a typical example of two normalized circuits and a final-use one. We can get circuit values between the two normalized circuits by appropriate modifications of the simple rules of Fig. 3-2.

To raise the impedance of a circuit, proportionately raise the impedance of everything. To increase the impedance by 5, *multiply* all resistors by 5. But, remember that capacitive reactance is inversely proportional to capacitor size; the larger the *capacitor,* the *lower* its impedance. So, to raise the impedance of a capacitor by 5, *divide* its value by 5. To scale impedance, multiply the resistors and divide the capacitors by the desired value.

When moving the center frequency of a circuit, you should look carefully at the particular filter circuit involved. Tuning guides will detail this later. In most of the filters of this book, very few components are used to affect the operating frequency. Most often they are two resistors or two capacitors for a second-order filter, and a single resistor and capacitor for a first-order network. It is usually very important to keep the *ratio* of all frequency-determining components constant. Usually, *both* resistors must be kept the same value or *both* capacitors must be kept the same value (or some other specified ratio of each other). Failure to keep this ratio will ruin the performance of the filter.

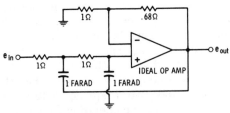

(A) Typical low-pass active filter normalized to 1 ohm and 1 radian per second. Use for *analysis*.

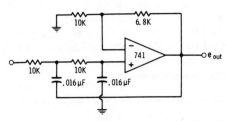

(B) Same circuit normalized to 10K ohms and 1-kHz cutoff. Use for *design*.

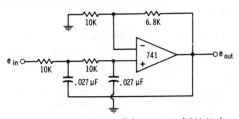

(C) Same circuit moved to final cutoff frequency of 588 Hz by scaling.

Fig. 3-1. Normalization and scaling techniques greatly simplify active-filter design.

To get *FROM* a 1-ohm, 1 radian-per-second ANALYSIS circuit *TO* a 1-kHz, 10K DESIGN circuit:

MULTIPLY all resistors by 10,000
DIVIDE all capacitors by 62.8 million, or 6.28×10^7.

To get *FROM* a 1-kHz, 10K DESIGN circuit *TO* a 1-radian-per-second, 1-ohm ANALYSIS circuit:

DIVIDE all resistors by 10,000
MULTIPLY all capacitors by 62.8 million, or 6.28×10^7.

Fig. 3-2. Normalizing rules

To raise the frequency of a circuit, *multiply* all frequency-determining resistors or all frequency-determining capacitors by the inverse ratio of the old frequency to the new one. Remember to keep the ratio of the resistors to a specified value and to keep the ratio of the capacitors to a specified value (often 1:1) at all times.

For instance, most second-order sections will have two frequency-determining resistors and two frequency-determining capacitors. Halve the resistors to double the operating frequency. Or, halve the capacitors to double the operating frequency. Do both and you will quadruple the cutoff frequency. Fig. 3-3 summarizes the scaling rules that let you move a 10K, 1-kHz filter to any desired cutoff frequency.

In a filter consisting of several cascaded sections, the impedance of any individual section can be changed to anything within reason, without affecting the response shape. On the other hand, *if the frequency of several cascaded sections is scaled, each section must be scaled by exactly the same amount.*

Always start with a 1-kHz, 10K impedance level circuit.

TO SCALE IMPEDANCE:

To change impedance, *multiply* all resistors and divide all capacitors by the new value, expressed in units of 10K. A 20K impedance level doubles all resistors and halves all capacitors. A 3.3K impedance level cuts all resistors by 1/3 and triples capacitor values.

Individual sections of a cascaded active filter can have their impedance levels individually scaled without altering the overall response shape.

TO SCALE FREQUENCY:

USING ONLY THE FREQUENCY-DETERMINING CAPACITORS:

Keep the ratio of both frequency-determining capacitors constant. *Double* the capacitors to *halve* the frequency, and vice versa. If the 1-kHz capacitor value is .016 μF, changing to 1600 pF will raise the frequency to 10 kHz. Similarly, a 0.16-μF capacitor will lower the frequency to 100 Hz.

USING ONLY THE FREQUENCY-DETERMINING RESISTORS:

Keep the ratio of both frequency-determining resistors constant. *Double* the resistance to *halve* the frequency. A 33K resistor lowers the frequency to one third of its initial value.

INDIVIDUAL SECTIONS OF A COMPOSITE ACTIVE FILTER CANNOT HAVE THEIR CUTOFF FREQUENCY CHANGED WITHOUT DRAMATICALLY CHANGING THE OVERALL RESPONSE SHAPE. If one section of a finalized filter is scaled in frequency, ALL sections must be scaled by the same amount.

Fig. 3-3. Scaling rules.

Scaling is best shown in going from the 1-ohm, 1-farad normalized-for-analysis circuit to the 10K, 0.16-μF, ready-for-use circuit. We might start by moving the frequency to 1 Hz. One hertz is 2π, or 6.28, radians per second, so a normalization to 1 Hz would take a 1-ohm resistor and a 1/6.28-farad capacitor (.16 farad). Next we can move to 1 kHz. We can do this by changing the capacitor. To increase the frequency by 1000, divide the capacitor by 1000. This gives a 160-μF capacitor and a 1-ohm resistor at a 1-kHz cutoff frequency.

Fnally, we can raise impedance. To do this, *multiply* the resistor by 10,000, getting 10K, and divide the capacitor by 10,000, getting .016 microfarad. In this way, the farad-sized capacitors used for analysis are readily reduced to more practical values when changing to the normal circuit frequencies and impedance levels.

Yet another normalization trick is used sometimes in this text. When you are analyzing the theory behind active filters, it is sometimes convenient to normalize a component to some handy value, perhaps setting an inductor to a value of "Q" henrys or "d" henrys or whatever. Once this is done, the analysis is very simple, and it is usually easy to scale components to final values later on. The same type of benefit can sometimes be had by forcing some point in a circuit to some convenient value—say one volt—and then letting everything else fall in place.

For analysis, work with 1 ohm and 1 radian per second. For synthesis, work with 1 kilohertz and 10K impedance levels.

To *raise* impedance, *multiply* all the resistors and *divide* all the capacitors. To *raise* frequency, *divide* the capacitors or the frequency-determining resistors. Always hold the frequency-determining resistors and the frequency-determining capacitors to the ratio called for by the filter. Remember, you can scale the impedance of any section at any time, but if you change frequency, the frequency of ALL cascaded sections must be changed identically.

FIRST-ORDER LOW-PASS SECTION

Fig. 3-4 shows the first-order low-pass section. An operational amplifier may be used as a voltage follower to isolate any load from the circuit. Optionally, it can provide a positive circuit gain, K, as shown. The op amp does not enter into any energy exchange or feedback in this circuit; it simply unloads a passive RC section.

The math involved is shown in Fig. 3-5, while the plots of amplitude and phase versus frequency appear as Figs. 3-6 and 3-7.

At very low frequencies, the capacitor does not load the resistor, giving unity gain and nearly zero degrees of phase shift. At very high frequencies, the capacitor shunts everything heavily to ground, resulting in heavy attenuation and a phase shift of nearly 90 degrees.

At a cutoff frequency of f = 1 for a normalized section, the resistance equals the capacitive reactance. The *vector* sum of the two acting as a voltage divider attenuates the output to 0.707 amplitude, or 3 decibels below the low-frequency value. This is a low-pass filter, with the passband being all the frequencies below the cutoff frequency and the stopband being all the frequencies above. The phase shift at cutoff is 45 degrees.

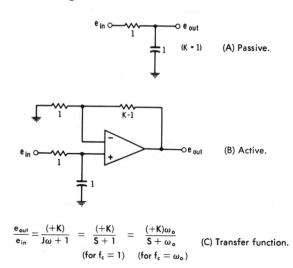

$$\frac{e_{out}}{e_{in}} = \frac{(+K)}{J\omega + 1} = \frac{(+K)}{S + 1} = \frac{(+K)\omega_o}{S + \omega_o}$$

(for $f_c = 1$)   (for $f_c = \omega_o$)   (C) Transfer function.

Fig. 3-4. First-order, low-pass sections.

If the resistor is 1 ohm and the capacitor is 1 farad, the cutoff frequency will be 1 radian per second. If the resistor is 10K and the capacitor is .016 microfarad, the cutoff frequency will be 1 kilohertz.

The slope of the response well above the cutoff frequency is −6 dB per octave. This means the amplitude is halved for every doubling of frequency. Since the highest power of frequency in this expression is unity, this is called a *first-order* section.

FIRST-ORDER HIGH-PASS SECTION

Fig. 3-8 shows the first-order high-pass section. It is simply an "inside out" version of the low-pass. Once again, the operational amplifier simply unloads the output and optionally provides gain. The math involved appears in Fig. 3-9. The amplitude and phase responses are shown in Figs. 3-10 and 3-11.

At high frequencies, the reactance of the capacitor is very low and the resistor does not significantly load the output. We have unity gain and a phase shift of nearly zero degrees. At very low frequen-

The first-order low-pass section.

Use the circuit of Fig. 3-4A. It is a voltage divider.

$$e_{out} = \frac{\text{impedance of capacitor}}{\text{total series impedance}} \times e_{in}$$

$$\frac{e_{out}}{e_{in}} = \frac{\dfrac{1}{j\omega C}}{R + \dfrac{1}{j\omega C}} = \frac{\dfrac{1}{j\omega C}}{\dfrac{j\omega RC + 1}{j\omega C}} = \frac{1}{j\omega RC + 1}$$

Where $\omega = 2\pi f$ and $j = \sqrt{-1}$

If we let $S = j\omega$ as a notation convenience

$$\frac{e_{out}}{e_{in}} = \frac{1}{j\omega + 1} = \frac{1}{S + 1}$$

1st-order low-pass

The amplitude will be $\dfrac{1}{\sqrt{1^2 + \omega^2}} = \dfrac{1}{\sqrt{1 + \omega^2}}$ or

expressed in dB of loss

$$\left| \frac{e_{out}}{e_{in}} \right| = 20 \log_{10}[1 + \omega^2]^{1/2}$$

amplitude response

And the phase is

$$\phi = -\tan^{-1}\frac{\omega}{1} = -\tan^{-1}\omega$$

phase angle

Fig. 3-5.

cies, the reactance of the capacitor is very high compared to the resistor, so we get heavy attenuation and a phase shift of nearly 90 degrees.

At the cutoff frequency, reactance and resistance are equal, and their vector sum drops the output 3 dB from its high-frequency value. The cutoff-frequency phase shift is 45 degrees.

The passband consists of all frequencies above cutoff, and the stopband consists of those frequencies below. This mirror-image re-

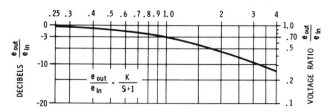

Fig. 3-6. Amplitude response—first-order low-pass section.

lationship between high-pass and low-pass is called *mathematical 1/f transformation*. Very handily, the design of a low-pass filter can usually be "turned inside out" to get an equivalent high-pass structure, simply by substituting resistors for capacitors and vice versa. You do have to make sure the op amp still gets its input base current from some dc return path to ground, but this is usually readily done.

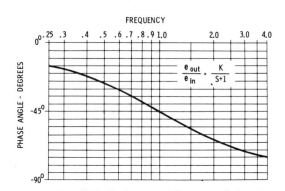

Fig. 3-7. Phase response—first-order low-pass section.

As with the first-order low-pass system, falloff is at a 6-dB per-octave rate, only in this case falloff is with decreasing rather than with increasing frequency. With either section, only the impedance level and the center frequency can be controlled.

SECOND-ORDER LOW-PASS SECTION

Fig. 3-12 shows the basic second-order low-pass sections. While there are many active circuits that will do this job, Fig. 3-12B is one that is simple and convenient for analysis. In theory, two resistors and two capacitors would be cascaded to get a second-order low-pass section. The trouble is that if this were done, the performance would be so poor that we probably could not use the results.

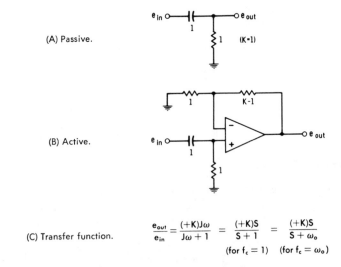

(A) Passive.

(B) Active.

(C) Transfer function.

$$\frac{e_{out}}{e_{in}} = \frac{(+K)J\omega}{J\omega + 1} = \frac{(+K)S}{S + 1} = \frac{(+K)S}{S + \omega_o}$$

(for $f_c = 1$)   (for $f_c = \omega_o$)

Fig. 3-8. First-order high-pass sections.

To build a useful second-order section, either you have to use an inductor and a capacitor together or else you must use an active circuit where an operational amplifier provides the equivalent of the energy-storage ability of the inductor. Instead of actually storing energy, the op amp can remove energy from the power supply and put it in the right place at the right time to get a transfer function like the one obtained by using an inductor and a capacitor.

Looking at things a bit differently, you could think of some active second-order low-pass sections as poorly performing two-resistor, two-capacitor, RC filters, each filter being followed by an op amp and a *feedback connection* that modifies or repairs the normally poor, highly damped, RC response and changes it into some more useful shape. Another possibility is a group of operational amplifiers which *compute* a response identical to an inductor-capacitor filter. Either way, what used to be energy stored in the inductor is now energy routed by the op amp from the supply and into the passive RC circuit.

Besides being essential in the energy feedback and response bolstering, the op amp also gives optional gain and a low output impedance that lets sections be cascaded without interaction.

The math behind this section is shown in Fig. 3-13, and the amplitude and phase response curves are shown in Figs. 3-14 and 3-15. This particular active circuit works by letting its capacitors have very little effect at low frequencies, which results in an essentially flat, very low frequency response. At very high frequencies, the capacitors separately shunt the signal to low-impedance points, one to

THE MATH BEHIND **The first-order high-pass section.**

Use the circuit of Fig. 3-8A. It is a voltage divider.

$$e_{out} = \frac{\text{impedance of resistor}}{\text{total series impedance}} \times e_{in}$$

$$\frac{e_{out}}{e_{in}} = \frac{R}{R + \frac{1}{j\omega C}} = \frac{R}{\frac{j\omega C + R}{j\omega}} = \frac{j\omega}{j\omega + 1}$$

where $\omega = 2\pi f$ and $j = \sqrt{-1}$.

If we let $S = j\omega$ as a notation convenience

$$\frac{e_{out}}{e_{in}} = \frac{d\omega}{j\omega + 1} = \frac{S}{S + 1}$$

1st order high-pass

The amplitude will be $\frac{\omega}{\sqrt{1^2 + \omega^2}} = \frac{\omega}{\sqrt{1 + \omega^2}}$

Expressed in dB of loss

$$\left|\frac{e_{out}}{e_{in}}\right| = 20 \log_{10} \left[\frac{\omega^2}{1 + \omega^2}\right]^{1/2}$$

amplitude response

And the phase is

$$\phi = -\tan^{-1} \frac{1}{\omega}$$

phase angle

Fig. 3-9.

ground and one to a highly attenuated signal output. This two-step shunting causes response at very high frequencies to fall off as the *square* of frequency. Hence the name, a *second-order* section.

The performance at very low and very high frequencies turns out to be identical to two cascaded and isolated RC sections, starting out flat and ending up falling at 12 dB per octave.

With the active section, things are much more interesting around the cutoff frequency. With the second-order active section, a new

FREQUENCY

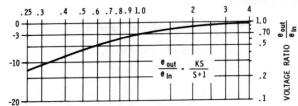

Fig. 3-10. Amplitude response—first-order high-pass section.

FREQUENCY

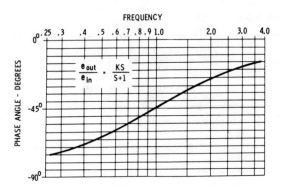

Fig. 3-11. Phase response—first-order high-pass section.

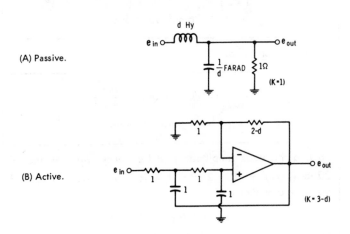

(A) Passive.

(B) Active.

(C) Transfer function.

$$\frac{e_{out}}{e_{in}} = \frac{+K}{dj\omega + (1 - \omega^2)} = \frac{+K}{S^2 + dS + 1} = \frac{+K\omega_o^2}{S^2 + d\omega_o S + \omega_o^2}$$

$$(f_c = 1) \qquad (f_c = \omega_o)$$

Fig. 3-12. Second-order low-pass sections.

The second-order low-pass section.

Use the circuit of Fig. 3-12A. It is a voltage divider with

$$e_{out} = \frac{\text{parallel impedance of R and C}}{\text{total series impedance}}\ e_{in}$$

$$\frac{e_{out}}{e_{in}} = \frac{K\dfrac{d}{d+j\omega}}{\dfrac{d}{d+j\omega}+d\omega} = \frac{K}{1+j\omega(d+j\omega)} = \frac{K}{-\omega^2+j\omega d+1}$$

or letting $S = j\omega$

$$\boxed{\frac{e_{out}}{e_{in}} = \frac{K}{(1-\omega^2)+j\omega d} = \frac{K}{S^2+dS+1}} \tag{A}$$
$$\text{second-order low-pass}$$

The amplitude response referred to K will be

$$\frac{1}{\sqrt{(1-\omega^2)^2+d^2\omega^2}},\ \text{or expanded and converted to decibels}$$

$$\boxed{\left|\frac{e_{out}}{e_{in}}\right| = 20\log_{10}\sqrt{\omega^4+(d^2-2)\omega^2+1}} \tag{B}$$
$$\text{amplitude response}$$

and the phase will be

$$\boxed{\phi = -\tan^{-1}\frac{d\omega}{1-\omega^2}} \tag{C}$$
$$\text{phase response}$$

To find the peak amplitude and frequency, we seek the minimum of $20\log_{10}\sqrt{\omega^4+(d^2-2)\omega^2+1}$ which for a given value of d means we want a minimum of

$$\omega^4+(d^2-2)\omega^2+1$$

We can do this by trial and error or by taking the derivative, following any basic calculus textbook. The derivative is

$$4\omega^3+2\omega(d^2-2)$$

and must be zero at a maximum or minimum

Fig. 3-13.

$$4\omega^2 + 2(d^2 - 2) = 0$$

$$2\omega^2 + (d^2 - 2) = 0$$

$$\omega_{max} = \sqrt{\frac{2 - d^2}{2}} = \sqrt{1 - \frac{d^2}{2}} \qquad (D)$$

peak frequency, if peak exists

A value of $d^2 = 2$ or $d = 1.414$ will have no peak and will be the flattest possible amplitude.

Putting the frequency of (D) into (B) gives us

$$\text{peak } \frac{e_{out}}{e_{in}} = 20 \log_{10} \sqrt{\left(\frac{2 - d^2}{2}\right)^2 + \frac{(d^2 - 2)(2 - d^2)}{2} + 1}$$

$$= 20 \log_{10} \frac{1}{2} \sqrt{4 - (2 - d^2)^2}$$

$$\frac{e_{out}}{e_{in}} = 20 \log_{10} \left[\frac{d\sqrt{4 - d^2}}{2}\right] \qquad (E)$$

peak amplitude, if peak exists

(E) will be negative since we are using decibels of *loss*.

Fig. 3-13—continued.

parameter called the *damping* (d) generates a family of curves near the cutoff frequency. These curves range from $d = 2$ (obtained by isolating and cascading two passive RC sections) to a best-time-delay curve of $d = 1.73$ and a flattest-amplitude curve of $d = 1.41$,

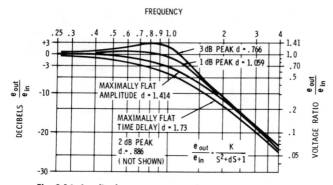

Fig. 3-14. Amplitude response—second-order low-pass section.

through lower and lower values of d that provide a more and more peaked response. Finally, at d = 0, you get an infinite-peak response, or an oscillator.

The second-order section is more obviously a low-pass filter since its passband and stopband areas are much better defined than those of the first-order low-pass. To actually build a second-order low-pass, a simple frequency translation, or scaling, is needed to make the desired curve have its −3-dB response curve correspond to a desired cutoff frequency. This text will always refer to the −3-dB cutoff frequency as 3 dB below the *peak* value of the response shape.

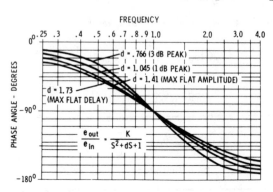

Fig. 3-15. Phase response—second-order low-pass section.

With the passive LC circuit, d is changed by changing the ratio of the inductor to the capacitor. As d gets smaller, the inductor gets smaller and the capacitor gets larger. This reduces the loading of the output resistor and gives a more peaked response.

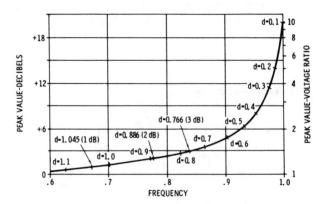

Fig. 3-16. Peak amplitude and frequency, second-order, low-pass section with low damping.

With the active-filter circuit shown, damping is lowered by providing more gain from the operational amplifier. As the gain goes up, proportionately more signal is fed back near the cutoff frequency, and the response becomes more and more peaked. If too much signal were fed back, we would have an oscillator. Fortunately, most of the d values needed in low-pass active filters are far removed from gain values that could cause oscillation. In addition, gain values are usually quite easy to stably set as a ratio of two passive resistors, thus minimizing variations that might tend toward oscillation.

A second-order filter is rarely designed with more than a 3-dB peaking (d = 0.766), but when we cascade sections for fancier response shapes, lower values of d are often called for. Fig. 3-16 shows the peak amplitude and the peak frequency obtained with lower damping values.

SECOND-ORDER BANDPASS SECTION

The series RLC circuit of Fig. 3-17A and one possible active equivalent (Fig. 3-17B) give us a simple resonant pole that is equivalent to a second-order bandpass response. We can use one of these by it-

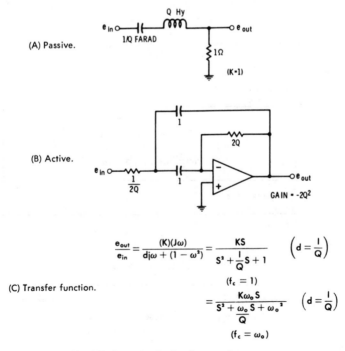

(A) Passive.

(B) Active.

(C) Transfer function.

$$\frac{e_{out}}{e_{in}} = \frac{(K)(J\omega)}{dj\omega + (1 - \omega^2)} = \frac{KS}{S^2 + \frac{1}{Q}S + 1} \qquad \left(d = \frac{1}{Q}\right)$$

$$(f_c = 1)$$

$$= \frac{K\omega_o S}{S^2 + \frac{\omega_o}{Q}S + \omega_o{}^2} \qquad \left(d = \frac{1}{Q}\right)$$

$$(f_c = \omega_o)$$

Fig. 3-17. Second-order bandpass sections.

THE MATH BEHIND **A second-order bandpass section.**

Use the circuit of Fig. 3-17A. It is a voltage divider with

$$e_{out} = \frac{\text{impedance of resistor}}{\text{total series impedance}} \times e_{in}$$

$$\frac{e_{out}}{e_{in}} = \frac{K}{1 + j\left[Q\omega - \frac{Q}{\omega}\right]} = \frac{K}{1 + jQ\left[\frac{\omega^2 - 1}{\omega}\right]} = \frac{(K)j\omega}{j\omega + Q(1 - \omega^2)}$$

$$= \left(\frac{K}{Q}\right)\frac{j\omega}{\omega^2 - 1 + j\frac{\omega}{Q}} \text{ or letting } K' = \frac{K}{Q} \text{ and } S = j\omega$$

$$\frac{e_{out}}{e_{in}} = K'\frac{j\omega}{(\omega^2 - 1) + j\frac{\omega}{Q}} = \frac{K'S}{S^2 + \frac{1}{Q}S + 1} \qquad \text{(A)}$$

second-order bandpass

$$S = \frac{1}{d}$$

The amplitude will be $\dfrac{1}{\sqrt{1^2 + Q^2\left[\frac{\omega^2 - 1}{\omega}\right]^2}}$, or

Relating to decibels of loss

$$\left|\frac{e_{out}}{e_{in}}\right| = 20 \log_{10}\sqrt{1 + Q^2\left[\frac{\omega^2 - 1}{\omega}\right]^2} \qquad \text{(B)}$$

amplitude response

And the phase will be

$$\phi = -\tan^{-1} Q^2\left[\frac{\omega^2 - 1}{\omega}\right]$$

phase response

Since $\omega = 2\pi f$, either $\omega = 1$ or $f = 1$ may be used interchangeably in any of these expressions.

Fig. 3-18.

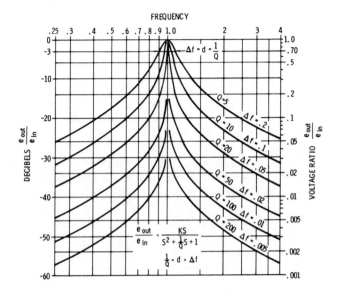

FREQUENCY

$$\frac{e_{out}}{e_{in}} = \frac{KS}{S^2 + \frac{1}{Q}S + 1}$$

$$\frac{1}{Q} = d = \Delta f$$

Fig. 3-19. Amplitude response—second-order bandpass section.

self, or we can cascade two, three, or more sections. By carefully staggering the frequencies of the cascaded poles slightly, we can control the overall response shape.

The math behind this section appears in Fig. 3-18, and the amplitude and phase response curves are shown as Figs. 3-19 and 3-20. Unlike the low-pass section, the peak value always occurs at the resonant frequency, or at $f_c = 1$ for a normalized filter. This can, of course, be scaled in frequency as needed.

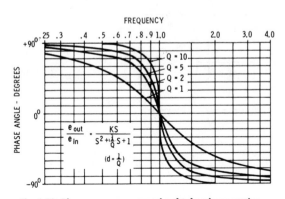

FREQUENCY

$$\frac{e_{out}}{e_{in}} = \frac{KS}{S^2 + \frac{1}{Q}S + 1}$$

$$(d = \frac{1}{Q})$$

Fig. 3-20. Phase response—second-order bandpass section.

At very low frequencies, the equivalent circuit is that of a series capacitor, so the low-frequency behavior is a rising, *first-order* response of 6 dB per octave. At the *resonant frequency,* the inductor, capacitor, and resistor interact to give a very peaked response. At resonance, the reactances cancel and the result is unity gain and zero degrees of phase shift.

At very high frequencies, the response is that of a single series inductor, a *first-order* response falling at a rate of 6 dB per octave.

The damping controls the peakedness of the response, just as it did with the low-pass section. The usual range of damping values called for involves extremely low values of d.

Because of this, it is convenient to use the inverse of the damping, called Q. Q is $1/d$. Q is also the bandwidth to the -3-dB points for a normalized $f_c = 1$ response. A damping value of .05 corresponds to a Q of 20.

The peak buildup of a single second-order bandpass section rises to "Q" above the point where the 6 dB per octave low- and high-frequency extended slopes would cross. Peak value is unity for the passive circuit and can be controlled in various active bandpass circuits.

The phase response starts out at a capacitive 90 degrees, swings through zero at resonance, and ends up as an inductive 90 degrees at very high frequencies.

Note that as you go away from the resonance frequency the band-pass response curves fall far less rapidly than at the start. Rejection values far from resonance cannot be predicted by continuing the

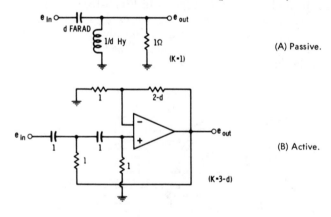

(A) Passive.

(B) Active.

$$\frac{e_{out}}{e_{in}} = \frac{(-1)(K)(\omega)^2}{dj\omega + (1 - \omega^2)} = \frac{(+K)S^2}{S^2 + dS + 1} = \frac{(+K)S^2}{S^2 + d\omega_o S + \omega_o^2}$$
$$(f_c = 1) \qquad (f_c = \omega_o)$$

(C) Transfer function.

Fig. 3-21. Second-order high-pass sections.

curves in a straight line near the -3-dB cutoff points. The important thing to note is that *no matter how high the Q of the bandpass pole, the low- and high-frequency falloff rates will only be 6 dB per octave per section.* All that the high Q gives is a narrower bandwidth to the -3-dB upper and lower cutoff frequencies and a more peaked response. Chapter 5 will give complete response curves on this.

THE MATH BEHIND   A second-order high-pass section.

Use the circuit of Fig. 3-21A. It is a voltage divider with

$$e_{out} = \frac{\text{parallel impedance of R and L}}{\text{total series impedance}} \times e_{in}$$

$$\frac{e_{out}}{e_{in}} = \frac{\dfrac{j\omega}{d}\,K}{\dfrac{j\omega}{d} + 1 + \dfrac{1}{j\omega d}} = \frac{Kj\omega}{j\omega + d + \dfrac{1}{j\omega}} = \frac{K(-\omega^2)}{1 - \omega^2 + j\omega d}$$

or, letting $S = j\omega$

$$\frac{e_{out}}{e_{in}} = K\frac{-\omega^2}{1 - \omega^2 + j\omega d} = \frac{KS^2}{S^2 + dS + 1} \qquad \text{(A)}$$

second-order high-pass

The amplitude response is given by

$$\left|\frac{e_{out}}{e_{in}}\right| = 20 \log_{10}\sqrt{\left(\frac{1}{\omega^4}\right) + \frac{d^2 - 2}{\omega^2} + 1} \qquad \text{(B)}$$

amplitude response

and the phase response is

$$\phi = \tan^{-1}\frac{d/\omega}{1 - \dfrac{1}{\omega^2}} \qquad \text{(C)}$$

phase response

(B) and (C) may be obtained by manipulating (A) or by taking the low-pass results (Fig. 3-14) and letting $f' = 1/f$ or $\omega' = 1/\omega$

Fig. 3-22.

SECOND-ORDER HIGH-PASS SECTION

The high-pass response is simply an inside-out, second-order low-pass one. The circuit for this section appears in Fig. 3-21, and the math and response plots are shown in Figs. 3-22, 3-23, and 3-24. This time, the high-frequency performance is flat and the low-frequency falloff drops at a 12-dB-per-octave rate, cutting to one-fourth for each halving in frequency.

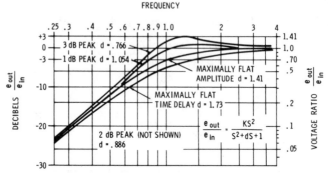

Fig. 3-23. Amplitude response—second-order high-pass section.

Near the cutoff frequency, once again there is a family of curves controlled by d, with identical d values used for comparable peaking. The peak frequencies are inverses of the low-pass; response curves appear in Fig. 3-25.

This high-pass response curve is based on the assumption that the op amp has a frequency response well beyond the maximum frequency of interest. In reality, there is no such thing as an active high-pass filter, for the falling response of the amplifier sets an upper limit to frequencies that can be passed.

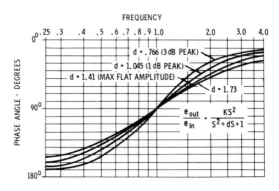

Fig. 3-24. Phase response—second-order high-pass section.

The phase of the second-order high-pass filter starts out at a value near zero at very high frequencies and ends up very near 180 degrees for very low frequencies. The phase shift of a second-order *low-pass* filter is −90 degrees at cutoff. The phase shift of a second-order *high-pass* is +90 degrees at cutoff. Note that these two responses can *can-*

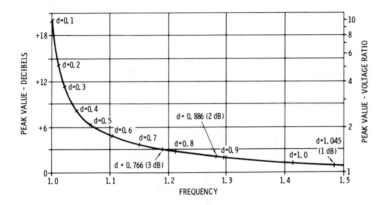

Fig. 3-25. Peak amplitude and frequency, second-order high-pass section with low damping.

cel each other out if you *sum* them together, giving a null at the cutoff frequency.

OTHER SECOND-ORDER RESPONSES

The remaining second-order responses are obtained by summing in different ways the three we already have. For instance, if we sum a second-order high-pass and a low-pass response, we get complete cancellation at the cutoff frequency, giving a *bandstop* response. A partial summing still gives a cancellation, only at a different frequency with different response amplitudes above and below cancellation. This useful effect applies to *elliptical,* or *Cauer,* filters; more details on this in Chapter 9. A summing of low-pass, bandpass, and high-pass together with inversion on the bandpass can give an *all-pass* response, useful where we want a constant amplitude but a controlled, varying phase response. Important applications are in equalizers and compensation networks. Combined bandpass–high-pass and bandpass–low-pass responses can be used in *formant* filters in electronic music applications. More details on these concepts appear in Chapter 10. Once we have the three basic second-order responses, we can easily combine them for these more specialized uses.

For a cutoff frequency of 1

	First Order	Second Order
Low-Pass	$\dfrac{K}{S+1}$	$\dfrac{K}{S^2+dS+1}$
Bandpass	(None)	$\dfrac{KS}{S^2+\dfrac{1}{Q}S+1}$ $(d=1/Q)$
High-Pass	$\dfrac{KS}{S+1}$	$\dfrac{KS^2}{S^2+dS+1}$

For a cutoff frequency of ω_o

	First Order	Second Order
Low-Pass	$\dfrac{K\omega_o}{S+\omega_o}$	$\dfrac{K\omega_o^2}{S^2+d\omega_o S+\omega_o^2}$
Bandpass	(None)	$\dfrac{K\omega_o}{S^2+\dfrac{\omega_o}{Q}S+\omega_o^2}$ $(d=1/Q)$
High-Pass	$\dfrac{KS}{S+\omega_o}$	$\dfrac{KS^2}{S^2+d\omega_o S+\omega_o^2}$

Fig. 3-26. The basic building-block transfer function.

K AND S

So far, not much has been said about the gain value, K, and the S that crops up in all the transfer functions.

K is simply a gain value, denoting how big the gain is for the entire response. K is usually something less than unity for passive sections. With active filters, K is controllable either to get a desired gain or to set a desired response into a particular active section. K affects all frequencies identically. Its value can be changed by adding gain or resistive padding either before, during, or after active filtering.

In this book, S is simply a notation convenience for any place we normally would write a $j\omega$ or jf expression. S is a *complex variable* that becomes extremely important if advanced analysis techniques are used in active filter design. We use S for two reasons: first, to make the notation simple and compact, and second, to get all the math into a standard form. S has both amplitude and phase associated with it. If $S = j\omega$ or jf, a single S has a phase of 90 degrees and S^2 has a phase of 180 degrees, or simply a sign change from $+$ to $-$.

SUMMING THINGS UP

Many interesting and useful filters can be built by properly cascading the five elemental filter sections. These are the first-order high-pass and low-pass, and the second-order low-pass, bandpass, and high-pass. Fig. 3-26 summarizes the transfer functions of these elemental sections, using S notation.

What remains now for the next two chapters is to find a way of specifying how many sections to use and what the damping, Q, and relative center frequencies are to be for a given response. Note that you cannot simply cascade identical sections and still end up with something reasonable in the way of response. For instance, a $d = 1.4$, second-order low-pass filter has the flattest amplitude of any you can build. Cascade three of these and, sure enough, you get a sixth-order filter. The trouble is that instead of being flat out to a -3-dB cutoff frequency, the old -3-dB point is now a -9-dB one, and you end up with a very drooped response. The trick is to find the mathematical second-order *factors* that, *when cascaded* properly, multiply together to get the final desired response.

CHAPTER 4

Low-Pass and High-Pass Filter Responses

In the last chapter, five elemental building blocks were covered: the first-order low-pass and high-pass sections, and the second-order low-pass, bandpass, and high-pass sections. While these are somewhat useful by themselves as simple active filters, they are much better performers when they are combined by suitable cascading to build higher-order filters. This chapter will show how to choose the proper values of damping and relative cutoff frequencies for each section, along with the needed component tolerances for seven basic and useful filter-response shapes, both high-pass and low-pass. The bandpass response is just enough different that Chapter 5 has been reserved for it.

ORDER

The *order* of a filter is given by the highest power of frequency, or radian frequency, that appears under the e_{out}/e_{in} *transfer function*. For instance, in a first-order low-pass section, the highest power of ω, or f, is 1; it is a first-order section, and the best *ultimate* falloff that can be expected at very high frequencies is $1/f$ or a halving of response as the frequency is doubled. This is equivalent to a slope of 6 dB per octave. The second-order lowpass filter has an S^2, an f^2, or an ω^2 under the transfer function. For values where f is large, f^2 is much larger, and the response eventually falls off as the square of frequency, going down by a factor of 4 as the frequency doubles. This is equal to a slope of 12 dB per octave.

As the order of a filter increases, the highest power of frequency under the transfer function increases, and the ultimate response fall-off versus frequency gets better. The rate is 6N dB per octave, where N is the order.

Fig. 4-1 shows the ultimate rate of falloff for filters of orders one through six. A sixth-order low-pass filter has an ultimate response

Filter Order	Ultimate Slope Low-Pass	Ultimate Slopes Bandpass*	Ultimate Slope High-Pass
1	—6 dB/octave	——————	+6 dB/octave
2	—12 dB/octave	±6 dB/octave	+12 dB/octave
3	—18 dB/octave	——————	+18 dB/octave
4	—24 dB/octave	±12 dB/octave	+24 dB/octave
5	—30 dB/octave	——————	+30 dB/octave
6	—36 dB/octave	±18 dB/octave	+36 dB/octave

(*See Chapter 5—Bandpass filters are normally even-order only.)

Fig. 4-1. How the order of a filter sets the ultimate rejection slopes.

falloff of 36 dB per octave. We can build filters of orders one through six by properly cascading first- and second-order sections (Fig. 4-2). Each section is carefully chosen to be a *factor* of the overall response shape desired; the sections cascaded are rarely identical, and the response shape of the individual sections normally appears wildly different from the final response shape. Further, the cutoff frequencies of each individual section may also be very much different from the final overall cutoff frequency desired.

Bandpass filters are usually limited to second, fourth, sixth, and higher-order responses, as will be shown in the next chapter.

SELECTING A SHAPE

To improve the response of a filter, we can increase the order, as increasing the order increases the ultimate rate of cutoff well away from the cutoff frequency. Can we do anything else?

With a first-order filter, the answer is no, but with higher orders there are new things that can be adjusted. With a second-order filter, we can adjust the damping and thus the response shape near the cutoff frequency. With a third-order filter consisting of a cascaded first- and second-order section, we can adjust the damping and the ratio of the cutoff frequencies. With a fourth-order filter, we have two damping values and a frequency ratio to adjust, and so on.

So, while we are limited to a flat low-frequency response and an ultimate falloff rate of 6N dB per octave for a filter of a particular order, we usually can control the shape around the cutoff frequency to get a particular response.

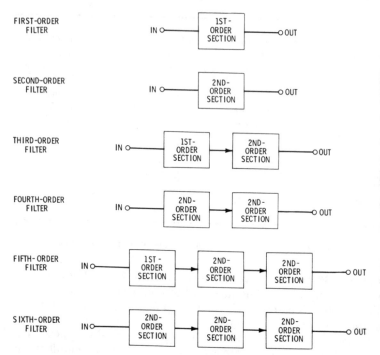

Fig. 4-2. Higher-order high-pass and low-pass filters are built by cascading first- and second-order active sections. Adjustment of frequency and damping of each section gives overall desired response shape.

The response shape you go after depends on what is important to you. For any property you consider important in a filter, you can find a mathematician who will dream up a wild set of functions to optimize it—at the expense of everything else. For example, you might consider the best possible time delay to be the most important. This would give you a filter with excellent transient response—but rotten attenuation skirts. Or, you might want to optimize the flattest possible passband. This is often a very useful compromise. Or, you might ask for a faster falloff outside the passband, but you will have to accept lumps (ripple) in the passband with it. Finally, you might want fast falloff to be optimized at the cost of everything else—even continuously increasing attenuation of the frequency response in the stopband. (This is an advanced filter technique requiring an *elliptical* filter—see Chapter 9.)

But, regardless of what property you optimize and regardless of how complex the math theory behind it is, when all is said and done, you simply have a stack of first- and second-order building blocks. On these you can control the damping and the frequency—no more, no less. All that the fancy math does is lead you in the right direction. You can get to the same place by trial and error. What you always end up with is a list of damping and frequency values.

For instance, on a second-order filter, if we let d = 1.73, we get a filter that has the best possible time-delay response but with a very gradual falloff and a drooping passband. If we set d to 1.41, we get a passband that is the flattest obtainable, with a moderately poorer transient and time-delay response. With d = 1.045, we get a 1-dB hump in the passband but a sharper *initial* falloff outside the passband. At d = 0.886, we get a 2-dB hump and a still faster falloff. At d = 0.766, we get a 3-dB hump and an initial falloff that is better yet. While all the initial falloff has been improving, the transient and overshoot response have been getting progressively worse. Lower d values will give too much of a good thing, and the passband will become too peaked to use in normal applications.

The problem, then, is to simply define several useful shapes for any order and normalize them to force all their 3-dB-below-peak cutoff frequencies at f = 1. Regardless of how peaked the filter is, we will define the cutoff frequency to be 3 dB below peak on the way out of the passband.

Once these curves for f = 1 are obtained, it is a simple matter of scaling to adjust the final cutoff frequency to anything you like.

You can use the curves of this chapter in several different ways. They can show the options in filter shapes for a particular response problem, given what has to be passed and what has to be rejected and by how much. They can show how a particular selection of filter order and shape will perform versus frequency. Most important, the charts that accompany the curves will tell how many basic first- and second-order sections are needed to do the job, along with the exact locations of their individual relative cutoff frequencies, their damping values, and the accuracy that must be maintained to realize the desired shape.

Note as you go through this chapter that we *never* use identical cascaded sections—at the least their damping values will differ. Except for the flattest-amplitude-filter shape option, the frequency values of each section will also differ by a specified amount. Each section must be a mathematical factor of the composite response we are after and not just the result of a bunch of sections stuck together.

A low-pass transfer function is usually expressed as a *polynomial* in "S." For instance, a fifth-order filter might have a transfer function of

$$\frac{e_{out}}{e_{in}} = f(S) = \frac{1}{S^5 + aS^4 + bS^3 + cS^2 + dS + 1} \qquad (1)$$

Remembering that we can let $S = j\omega$, S^5 corresponds to a fifth power of ω or frequency. At very low frequencies ω is small and powers of ω are even smaller, so the response is simply $\frac{1}{1}$ or 1. At very high frequencies ω^5 is very large, so the response falls off at $\frac{1}{\omega^5}$ or 30 dB/octave. We can get any response shape we might reasonably want near $\omega = 1$ by a correct choice of a, b, c, and d. Generally a, b, c, and d are complex numbers.

For reasonableness and ease of active-filter design, we can *factor* the polynomial into first- and second-order sections. For instance

$$f(S) = \frac{1}{S^5 + aS^4 + bS^3 + cS^2 + dS + 1}$$

$$= \frac{1}{S^2 + vS + w} \times \frac{1}{S^2 + xS + y} \times \frac{1}{S + z}$$

And our v, w, x, y, and z are uniquely related to a, b, c, and d by multiplying them out and solving simultaneous equations. Once this is done, each first-order section will simply specify its own frequency, while each second-order section will specify its frequency and damping.

To get a desirable response characteristic, we can use trial and error adjustment of the frequency and damping values of the *factored* sections. Or we can start with a polynomial that has known, useful properties and factor it to get the desired damping and frequency values.

There are lots of good polynomials to use. A polynomial that gives the best possible time delay is called a *Bessell* polynomial. A *Butterworth* polynomial has the flattest amplitude, and so on. These and other polynomials appear in factored form in *Network Analysis and Synthesis,* L. Weinberg, and *Operational Amplifiers—Design and Applications,* G. E. Tobey, listed in the references. All we have done in the curves that follow is use these factors, adjusting them slightly in frequency to get all the −3-dB response points to correspond to the cutoff frequency.

The actual response plots are most easily made by finding the response of the individual sections and then simply adding the decibel loss

Fig. 4-3.

and phase angle of each section. Note that adding decibels is the same as multiplying the actual numeric loss of cascaded sections.

The loss of a first-order section is found by finding the amplitude of $\frac{1}{S+1}$ or, letting $S = j\omega$, $\frac{1}{j\omega + 1}$. The amplitude will be $\frac{1}{\sqrt{\omega^2 + 1}}$, or expressed as decibels of *loss*

$$\text{Amplitude of first-order section} = 20 \, \log_{10}[\omega^2 + 1]^{1/2}$$
$$\text{in decibels}$$

and the phase is simply

$$\text{Phase angle of first-order section} = \tan^{-1}\omega$$

A second-order section amplitude response is found from

$$\frac{1}{S^2 + dS + 1};$$ or, letting $S = j\omega$, the response becomes

$$\frac{1}{(1 - \omega^2) + jd\omega}.$$

The amplitude is then

$$\frac{1}{\sqrt{(1 - \omega^2)^2 + (d\omega)^2}} = \frac{1}{\sqrt{\omega^4 + [d^2 - 2]\omega^2 + 1}}$$

or expressed as decibels of loss

$$\text{Amplitude of second-order section} =$$
$$20 \, \log_{10}[\omega^4 + (d^2 - 2)\omega^2 + 1]^{1/2}$$
$$\text{in decibels}$$

and the phase is

$$\text{Phase of second-order section} = \tan^{-1}\frac{d\omega}{1 - \omega^2}$$

Usually ω or f will differ from the desired filter-cutoff frequency. A simple scaling of frequency is then needed, forming a new f' or ω' by *dividing* by the design frequencies. For instance, if the section is supposed to have a design frequency of $.834f$, the response of that section at f will be the same as a normalized ($f = 1$) section at a frequency of $\frac{1}{.834} = f' = 1.199$.

Fig. 4-3—continued.

LOW-PASS RESPONSE CURVES

The math behind the low-pass response curves appears in Fig. 4-3, and the curves themselves are shown in Figs. 4-4 through 4-9. Along with each curve is a chart showing the frequency and damping values needed for each cascaded first- or second-order section that is used to make up the filter.

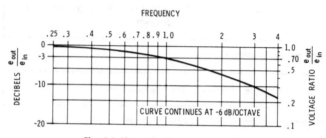

Fig. 4-4. First-order low-pass response.

The charts show a total of seven possible different filter-shape options, while only four responses are shown on the graphs. The missing response curves are simply read halfway between the existing curves. All of the response options assume that there is to be no ripple in the stopband and that once it is started, attenuation continues to increase without limit with increasing frequency.

The shape options trade off smoothness of response against rapidity of falloff. Their only real difference is in the specification of the relative cutoff frequency and the damping values of each cascaded section. The responses available are:

Best-Time-Delay Filter—Sometimes called a *Bessel* filter. This one has the best possible time delay and overshoot response, but it has a droopy passband and very gradual initial falloff.

Compromise Filter—Often called a *Paynter* or *transitional Thompson-Butterworth* filter. It has a somewhat flatter passband and initially falls off moderately faster than the best-time-delay filter, with only moderately poorer overshoot characteristics.

Flattest-Amplitude Filter—This is the *Butterworth* filter and has the flattest passband you can possibly provide combined with a moderately fast initial falloff and reasonable overshoot. The overshoot characteristics appear in Fig. 4-10. *The Butterworth is often the best overall filter choice.* It also has a characteristic that sets all cascaded sections to the same frequency, which makes voltage control and other wide-range tuning somewhat easier.

Slight-Dips Filter—This is the first of the *Chebyshev filters.* It has a slight peaking or ripple in the passband, a fast initial falloff, and a

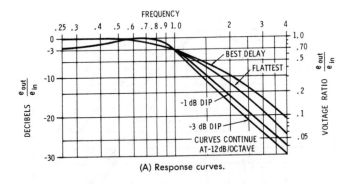

(A) Response curves.

A second-order filter is built with a single second-order section. Its ultimate attention rate is —12 dB/octave.

For a cutoff (—3 dB) frequency of f, the section parameters are:

Filter Type	Second-Order Section	
	Frequency	Damping
Best Delay	1.274 f	1.732
Compromise	1.128 f	1.564
Flattest Amp	1.000 f	1.414
Slight Dip	0.929 f	1.216
1-Decibel Dip	0.863 f	1.045
2-Decibel Dip	0.852 f	0.895
3-Decibel Dip	0.841 f	0.767

Zero frequency attenuation is 0 decibels for first four filter types, —1 dB for 1-dB dip, —2 dB for 2-dB dip, and —3 dB for 3-dB dip filter types.
NOTE—Values on this chart valid *only* for second-order filters. See other charts for suitable values when sections are cascaded.

(B) Section values.

Fig. 4-5. Second-order low-pass filters.

transient response only slightly worse than the flattest-amplitude filter. The ripple depends on the order and varies from 0.3 dB for the second-order response down to .01 dB at the sixth-order.

One-dB-Dips Filter—This is another Chebyshev filter. It has 1 dB of passband ripple. The ripple peaks and troughs are constant in amplitude, but you get more of them as the order increases. They

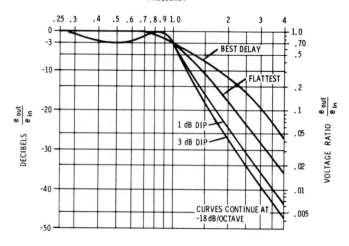

(A) Response curves.

A third-order filter needs a cascaded first- and second-order section. Its ultimate attenuation rate is —18 dB/octave.

For a cutoff (—3 dB) frequency of f, the section parameters are:

| Filter Type | Second-Order Section | | First-Order Section |
	Frequency	Damping	Frequency
Best Delay	1.454f	1.447	1.328f
Compromise	1.206f	1.203	1.152f
Flattest Amp	1.000f	1.000	1.000f
Slight Dips	0.954f	0.704	0.672f
1-Decibel Dips	0.911f	0.496	0.452f
2-Decibel Dips	0.913f	0.402	0.322f
3-Decibel Dips	0.916f	0.326	0.299f

Zero frequency attenuation is 0 decibels for all filter types.

(B) Section values.

Fig. 4-6. Third-order low-pass filters.

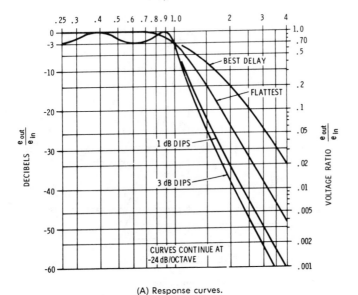

(A) Response curves.

A fourth-order filter needs two cascaded second-order sections. Its ultimate attenuation rate is —24 dB/octave.

For a cutoff (—3 dB) frequency of f, the section parameters are:

	First Section		Second Section	
Filter Type	Frequency	Damping	Frequency	Damping
Best Delay	1.436f	1.916	1.610f	1.241
Compromise	1.198f	1.881	1.269f	0.949
Flatest Amp	1.000f	1.848	1.000f	0.765
Slight Dip	0.709f	1.534	0.971f	0.463
1-Decibel Dip	0.502f	1.275	0.943f	0.281
2-Decibel Dip	0.466f	1.088	0.946f	0.224
3-Decibel Dip	0.443f	0.929	0.950f	0.179

Zero frequency attenuation is 0 dB for first four filter types, —1 dB for 1-dB dip, —2 dB for 2-dB dip, —3 dB for 3-dB dip filter types.

(B) Section values.

Fig. 4-7. Fourth-order low-pass filters.

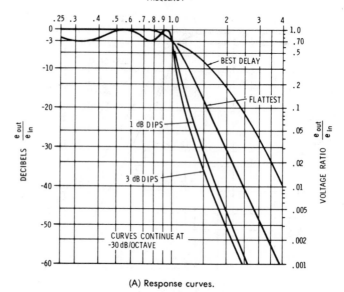

(A) Response curves.

A fifth-order filter needs two cascaded second-order sections cascaded with a single first-order section. Its ultimate attenuation rate is —30 dB/octave.

For a cutoff (—3 dB) frequency of f, the parameters of the sections are:

Filter Type	Second-Order 1st Section		Second-Order 2nd Section		First-Order Section
	Frequency	Damping	Frequency	Damping	Frequency
Best Delay	1.613f	1.775	1.819f	1.091	1.557f
Compromise	1.270f	1.695	1.348f	0.821	1.248f
Flattest Amp	1.000f	1.618	1.000f	0.618	1.000f
Slight Dips	0.796f	1.074	0.980f	0.334	0.529f
1-Decibel Dips	0.634f	0.714	0.961f	0.180	0.280f
2-Decibel Dips	0.624f	0.578	0.964f	0.142	0.223f
3-Decibel Dips	0.614f	0.468	0.967f	0.113	0.178f

Zero frequency attenuation is 0 dB for all filter types.

(B) Section values.

Fig. 4-8. Fifth-order low-pass filters.

tend to crowd together near the cutoff frequency, particularly when viewed on a log response plot.

Two-dB-Dips Filter—Another Chebyshev filter. The 2-dB ripple gives faster initial stopband falloff and progressively poorer transient and overshoot characteristics.

Three-dB-Dips Filter—This final Chebyshev filter offers the fastest initial falloff you can possibly get in a filter with acceptable passband lumps and continually increasing attenuation in the stopband.

In general, the damping values of the individual sections become less and less as you proceed from best-time-delay to 3-dB Chebyshev. This means that the sections become somewhat harder to build and more critical in tolerance as you move toward the Chebyshev end of the shape options, with the 3-dB-dips filter being the most critical of the seven options.

The passbands of only the flattest-amplitude and 3-dB-dips filters are shown in the response curves. Compromise and best-time-delay filters will have more droop than the flattest-amplitude curve, but they will be ripple free. The slight-dips passband and the 1- and 2-dB-dips passbands will fall between the two curves shown. All filters are 3 dB down from their peak value at the cutoff frequency $f = 1$.

Note that the "dc" or ultralow-frequency response depends on the order and shape. For all odd-order filters and for all even-order, ripple-free filters (best-time-delay, compromise, and flattest-amplitude), the "dc" value of filter attenuation is zero decibels.

For even-order, underdamped filters with passband ripple, the dc value of the attenuation is down by the passband ripple in the curves shown. This standardizes the peak height and the −3-dB cutoff frequencies for meaningful comparison. A simple gain adjustment elsewhere is easily made to allow for the overall gain constant.

Once again, note that all the curves have a −3-dB cutoff frequency in common that defines the boundary between passband and stopband, independent of ripple amplitude or amount of delay.

HIGH-PASS-FILTER CHARACTERISTICS

We can likewise generate a family of high-pass response curves, following the math of Fig. 4-11 and the corresponding curves of Figs. 4-12 through 4-17. Thanks to a process called *mathematical transformation by 1/f*, the high-pass filters are simply mirror images of their low-pass counterparts. One minor difference is that the low-pass best-time-delay filter is more properly called a *well-damped* filter in the high-pass case. Outside of this minor semantic detail, the responses are simply mirror images of each other, obtained from a 1/f frequency transformation.

While the high-pass curves shown are flat theoretically to infinite frequency, the actual circuits of Chapter 8 will impose an upper cutoff frequency. In reality, there is no such thing as an active high-pass filter. What we really have is a bandpass filter whose lower cutoff frequency is set by the designed-in high-pass filter and whose upper cutoff frequency is determined by the op amp in use. Normally, you place this upper cutoff frequency limit beyond the range of whatever signals you are filtering.

HOW ACCURATE?

How closely must we match these curves to the real-world damping and frequency values we get from active sections? This is called the *sensitivity* problem, and it rapidly leads to a lot of difficult math. One simple and highly useful way to estimate the degree of accuracy required by the frequency and damping values of cascaded sections is to assume that a 1-dB change in the worst possible position of the most sensitive section sets an upper limit to what we will accept as a reasonable shape variation. We then force the other sections to also meet this requirement.

The math involved is shown in Fig. 4-18, and the main results, rounded off to stock tolerance values, appear in Fig. 4-19.

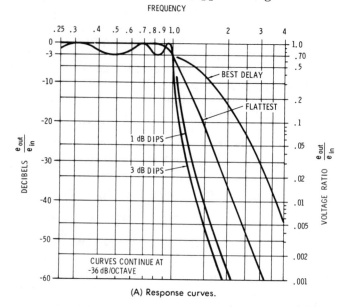

(A) Response curves.

Fig. 4-9. Sixth-order

A sixth-order filter needs three cascaded second-order sections. Its ultimate attenuation rate is —36 dB/octave.

For a cutoff (—3 dB) frequency of f, the parameters of the second-order sections are:

Filter Type	First Section Frequency	First Section Damping	Second Section Frequency	Second Section Damping	Third Section Frequency	Third Section Damping
Best Delay	1.609f	1.959	1.694f	1.636	1.910f	0.977
Compromise	1.268f	1.945	1.301f	1.521	1.382f	0.711
Flattest Amp	1.000f	1.932	1.000f	1.414	1.000f	0.518
Slight Dips	0.589f	1.593	0.856f	0.802	.988f	0.254
1-Decibel Dips	0.347f	1.314	0.733f	0.455	.977f	0.125
2-Decibel Dips	0.321f	1.121	0.727f	0.363	.976f	0.0989
3-Decibel Dips	0.298f	0.958	0.722f	0.289	.975f	0.0782

Zero frequency attenuation is 0 dB for first four filter types, —1 dB for 1-dB dip, —2 dB for 2-dB, and —3 dB for 3-dB dip filter types.

(B) Section values.

Order	Overshoot
2	5%
3	9%
4	11%
5	13%
6	15%

Fig. 4-10. Approximate overshoot of Butterworth (flattest-amplitude) low-pass filters to a sudden change of input.

THE MATH BEHIND **High-pass filter response.**

High-pass response analysis could be done just like the lowpass analysis of Fig. 4-3, for instance, starting with a polynomial such as

$$\frac{e_{out}}{e_{in}} = \frac{S^5}{S^5 + aS^4 + bS^3 + cS^2 + jS + 1}$$

And factoring it into first- and second-order sections:

$$\frac{e_{out}}{e_{in}} = \frac{S^2}{S^2 + vS + w} \times \frac{S^2}{S^2 + xS + y} \times \frac{S}{S + z}$$

Again relating a, b, c, and d to v, w, x, y, and z by multiplying out and solving simultaneous equations.

A much easier way to do the job is to let $S' = \frac{1}{S}$ or $f' = \frac{1}{f}$ to "inside out" the low-pass response. This is called a *frequency transformation* and is simply done by mirror-imaging the low-pass curves and design frequencies.

> To find a high-pass response equivalent to a low-pass one let $S' = \frac{1}{S}$, $\omega' = \frac{1}{\omega}$, or $f' = \frac{1}{f}$ everywhere in its expression, or use a mirror image graphically.

Fig. 4-11.

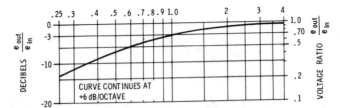

Fig. 4-12. First-order high-pass filter response.

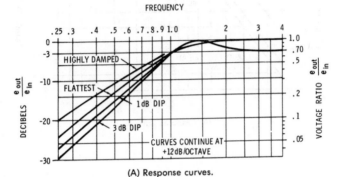

(A) Response curves.

A second-order filter is built with a single second-order section. Its ultimate attenuation rate is +12 dB/octave.

For a cutoff (—3 dB) frequency of f, the section parameters are:

| | Second-Order Section | |
Filter Type	Frequency	Damping
Highly Damped	0.785f	1.732
Compromise	0.887f	1.564
Flattest Amp	1.000f	1.414
Slight Dip	1.076f	1.216
1-Decibel Dip	1.159f	1.045
2-Decibel Dip	1.174f	0.895
3-Decibel Dip	1.189f	0.767

Very high frequency attenuation is 0 dB for first four filter types, —1 dB for 1-dB dip, —2 dB for 2-dB dip, and —3 dB for 3-dB dip filter types.

NOTE—Values on this chart valid *only* for second-order filters. See other charts for suitable values when sections are cascaded.

(B) Section values.

Fig. 4-13. Second-order high-pass filters.

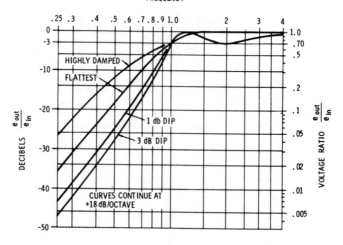

(A) Response curves.

A third-order filter needs a cascaded first- and second-order section. Its ultimate attenuation rate is +18 dB/octave.

For a cutoff (—3 dB) frequency of f, the section parameters are:

Filter Type	Second-Order Section		First-Order Section
	Frequency	Damping	Frequency
Highly Damped	0.688f	1.447	0.753f
Compromise	0.829f	1.203	0.868f
Flattest Amp	1.000f	1.000	1.000f
Slight Dip	1.048f	0.704	1.488f
1-Decibel Dip	1.098f	0.496	2.212f
2-Decibel Dip	1.095f	0.402	3.105f
3-Decibel Dip	1.092f	0.326	3.344f

Very high frequency attenuation is 0 dB for all filter types.

(B) Section values.

Fig. 4-14. Third-order high-pass filters.

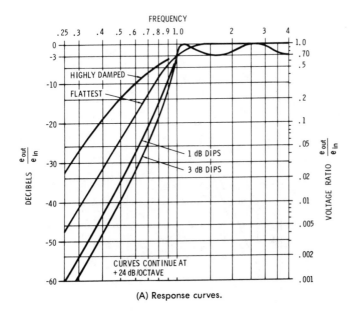

(A) Response curves.

A fourth-order filter needs two cascaded second-order sections. Its ultimate attenuation rate is $+24$ dB/octave.

For a cutoff (-3 dB) frequency of f, the section parameters are:

Filter Type	First Section		Second Section	
	Frequency	Damping	Frequency	Damping
Highly Damped	0.696f	1.916	0.621f	1.241
Compromise	0.834f	1.881	0.788f	0.949
Flattest Amp	1.000f	1.848	1.000f	0.765
Slight Dip	1.410f	1.534	1.029f	0.463
1-Decibel Dip	1.992f	1.275	1.060f	0.281
2-Decibel Dip	2.146f	1.088	1.057f	0.224
3-Decibel Dip	2.257f	0.929	1.053f	0.179

Very high frequency attenuation is 0 dB for first four filter types, -1 dB for 1-dB dip, -2 dB for 2-dB dip, and -3 dB for 3-dB dip filter types.

(B) Section values.

Fig. 4-15. Fourth-order high-pass filters.

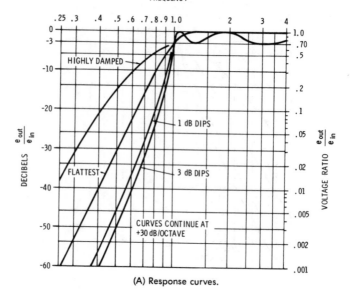

(A) Response curves.

A fifth-order filter needs two cascaded second-order sections cascaded with a single first-order section. Its ultimate attenuation rate is +30 dB/octave.

For a cutoff (−3 dB) frequency of f, the parameters of the sections are:

Filter Type	Second-Order 1st Section		Second-Order 2nd Section		First-Order Section
	Frequency	Damping	Frequency	Damping	Frequency
Highly Damped	0.620f	1.775	0.550f	1.091	0.642f
Compromise	0.787f	1.695	0.742f	0.821	0.801f
Flattest Amp	1.000f	1.618	1.000f	0.618	1.000f
Slight Dip	1.256f	1.074	1.020f	0.334	1.890f
1-Decibel Dip	1.577f	0.714	1.041f	0.180	3.571f
2-Decibel Dip	1.603f	0.578	1.037f	0.142	4.484f
3-Decibel Dip	1.629f	0.468	1.034f	0.113	5.618f

Very high frequency attenuation is 0 dB for all filter types.

(B) Section values.

Fig. 4-16. Fifth-order high-pass filters.

These results tell us that a 10% damping accuracy is good enough for all the filters of this chapter, while accuracies of 10% down to 1% are needed when the frequency of each section is set. Most of the circuits shown are easily handled with a 5% tolerance.

Often, an active-filter circuit uses two capacitors or two resistors together to determine frequency. This lets us achieve a nominal 1% frequency accuracy with components that are accurate to approximately 2%. It turns out that the majority of active filters are easily handled with fixed 5% components except for very special or critical needs.

On the other hand, we cannot be sloppy or radically out of tolerance. The sixth-order filters have damping values as low as 0.07. This response shape by itself has a peak of around 24 dB. Put this in the wrong place and you are bound to get a response that will be wildly wrong. The practical limitation is that you have to use the most-accurate components you can for a filter task. Such things as ganged, low-tolerance potentiometers for wide-frequency tuning should be avoided (see Chapter 9), as should individual tuning or damping adjustments that cover too wide a range.

USING THE CURVES

Here is how to use the data of this chapter:

1. From your filter problem, establish the cutoff frequency and the allowable response options. Also, choose some response-shape criterion that you would like to meet, in terms of a certain frequency, a certain amount above cutoff, to be attenuated by so many decibels.
2. Check through the curves to find what filter options you have. *Always try a flattest-amplitude filter first.* Remember that filters less damped than flattest-amplitude will fall off faster initially but will have ripple in the passband and a poorer transient response. Filters more damped than a flattest-amplitude filter will have good to excellent transient response but a very droopy passband and a poor initial falloff.
3. Read the frequency and damping values you need for each cascaded section and scale them to your particular cutoff frequency.
4. Determine the accuracy and the tolerance you need from Fig. 4-19.

Several examples appear in Fig. 4-20.

CAN WE DO BETTER?

Newcomers to the field of filter design may find some of these response shapes disappointing. Can we do any better?

Remember that each circuit shown optimizes *something*. The best-delay filter is the best and finest one you can possibly build. The flattest-amplitude filter is indeed the flattest. The 3-dB-dips filter drops off as fast as it possibly can, consistent with an increasing attenuation with changing frequency. *Each of these curves is the best possible design that you can achieve in filter work, for some feature of the filter.*

There are three additional things we can do if the curves of this chapter are inadequate:

Increase the Order—Higher-order filters will offer better responses, at the expense of more parts, tighter tolerance, and generally lower damping values. Parameter values can be found by trial and error or by consulting advanced filter-theory texts.

Go Elliptic—A certain type of filter that provides ripple in the stopband as well as in the passband gives you the fastest possible filter falloff and a null, or zero, just outside the passband—but the transient and overshoot performance is relatively poor, and frequencies far into the stopband are not attenuated much. Details on this type of filter are shown in Chapter 9. Circuits are relatively complex.

Consider Alternatives—If your circuit requirement cannot be achieved with the curves of this chapter, chances are that your specification is too restrictive or otherwise unrealistic. The over-

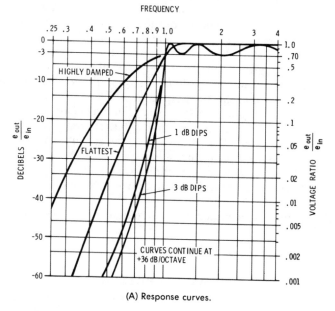

(A) Response curves.

Fig. 4-17. Sixth-order

A sixth-order filter needs three cascaded second-order sections. Its ultimate attenuation rate is $+36$ dB/octave.

For a cutoff (-3 dB) frequency of f, the parameters of the second-order sections are:

Filter Type	First Section		Second Section		Third Section	
	Frequency	Damping	Frequency	Damping	Frequency	Damping
Highly Damped	0.621f	1.959	0.590f	1.636	0.524f	0.977
Compromise	0.788f	1.945	0.768f	1.521	0.724f	0.711
Flattest Amp	1.000f	1.932	1.000f	1.414	1.000f	0.518
Slight Dip	1.697f	1.593	1.168f	0.802	1.012f	0.254
1-Decibel Dip	2.881f	1.314	1.364f	0.455	1.023f	0.125
2-Decibel Dip	3.115f	1.121	1.375f	0.363	1.025f	0.0989
3-Decibel Dip	3.356f	0.958	1.385f	0.289	1.026f	0.0782

Very high frequency attenuation is 0 dB for the first four filter types, -1 dB for 1-dB dip, -2 dB for 2-dB dip, and -3 dB for 3-dB dip filter types.

(B) Section values.

high-pass filters.

Tolerance and sensitivity analysis.

Finding the accuracy needed for the frequency and damping values of each section can be a very complicated and confusing task. This can be substantially simplified by assuming that a 1-dB change in the most underdamped (lowest d) section is an outer limit of acceptability, and then imposing this tolerance limit on the other sections.

This 1-dB shift can be calculated at the peak value of the section response, for it will usually be the most dramatic at that point. The amplitude response of a second-order section is

$$\frac{e_{out}}{e_{in}} = 20 \log_{10} [\omega^4 + (d^2 - 2)\omega^2 + 1]^{\frac{1}{2}}$$

and the peak frequency (d < 1.41) is

$$\omega_{max} = \sqrt{1 - \frac{d^2}{2}}$$

and the peak amplitude is

$$\frac{e_{out}}{e_{in}} = 20 \log_{10} \left[\frac{d\sqrt{4 - d^2}}{2} \right]$$

From Figs. 3-16 and 4-3 we can then select low values of d and shift them independently in frequency and damping to produce a 1-dB error. These values are then related to minimum d values needed for a given response and conservatively rounded off, resulting in the chart of Fig. 4-19.

Fig. 4-18.

whelming majority of practical filter problems can be handled with the circuits of this book. If yours is not one of them, consider some alternate technique, such as digital computer filters, sideband or multiplier modulation techniques, or phase-lock loops.

Filter Type	Order				
	2	3	4	5	6
Best Delay (LP) Highly Damped (HP)	±10%	±10%	±10%	±10%	±10%
Compromise	±10%	±10%	±10%	±10%	±10%
Flattest Amp	±10%	±10%	±10%	±10%	±10%
Slight Dips	±10%	±10%	±10%	±10%	± 5%
1-Decibel Dip	±10%	±10%	± 5%	± 5%	± 2%
2-Decibel Dip	±10%	± 5%	± 5%	± 2%	± 2%
3-Decibel Dip	±10%	± 5%	± 2%	± 2%	± 1%

Damping accuracy for any order, any filter—±10%.

Fig. 4-19. Tuning accuracy needed for low-pass or high-pass filters.

Using the curves; some **EXAMPLES**

A. A low-pass filter is to have its cutoff frequency at 1 kHz and reject all frequencies above 2 kHz by at least 24 dB. What type of filter can do this?

None of the first- or second-order filters offer enough attenuation. Third-order low-pass structures with 1-, 2-, or 3-dB dips in them will do the job. A fourth-order, maximally flat amplitude will just do the job, while a fourth-order "slight dips" filter offers a margin of safety. Even through sixth-order, a best-time-delay filter cannot offer this type of rejection. The best choice is probably the fourth-order flattest-amplitude filter.

B. A high-pass, fourth-order, 3-dB dips filter has its cutoff frequency at 200 Hz. What will the response be at 1000, 400, 200, 100, and 20 Hz?

Since 200 Hz is the cutoff frequency, the response here is —3 dB by definition. 400 Hz is 400/200, or twice the cutoff frequency. From Fig. 4-15A, the response is —2 dB and, of course, in the passband. 1000 Hz is 1000/200 or 5 times the cutoff frequency. It is off the curve, but is near the —3 dB "very high frequency" response point. At 100 Hz, the frequency is 100/200 or 0.5 times the cutoff frequency and the rejection from Fig. 4-15A will be —39 dB. Twenty hertz is only 0.1 times the cutoff frequency and is thus off the curve. By inspection, it is well below —60 dB.

If we want an exact value for 20 Hz, we can come up two octaves in frequency to 80 Hz. Eighty Hz is 80/200 times the cutoff frequency and the attenuation at 0.4 frequency is —47 dB. Since the curves continue at —24 dB per octave and since there are two octaves between 20 and 80 Hz, the 20-Hz attenuation is theoretically 47 + 24 + 24 = 95 dB. In the real world, attenuations greater than 60 dB are often masked by direct feedthrough, circuit strays, coupling, and so forth. Very high values of attenuation and rejection are only obtained with very careful circuit designs and using circuits with extreme dynamic ranges.

C. A third-order, 350 Hz, 1-dB-dip, low-pass filter is to be built. What are the damping and frequency values of the cascaded sections? How accurate do they have to be?

From Fig. 4-6B, we see that we need two sections, a first-order and a second-order one. The cutoff frequency of the first-order section is to be 0.452 times the design frequency or 0.452 × 350 = 158 Hz. The cutoff frequency of the second-order section is to be 0.911 times the cutoff frequency or 319 Hz, while its damping is read directly as 0.496.

From Fig. 4-19, a 10% accuracy on the damping and a 10% accuracy on frequency should be acceptable.

These values may then be taken to Chapter 6 for actual construction of the filter.

Fig. 4-20.

CHAPTER 5

Bandpass Filter Response

In this chapter, you will find out how to decide what the response shape of a bandpass filter is and how to properly pick the center frequency, the Q, and the frequency offset of cascaded filter sections.

The technique we will use is called *cascaded pole synthesis.* In it, you simply cascade one, two, or three active second-order bandpass circuits to build up an overall second-, fourth-, or sixth-order response shape. By carefully choosing the amount of staggering and the Q of the various sections, you can get a number of desirable shape factors. The advantages of this method are that it is extremely simple to use, requires no advanced math, and completely specifies the entire response of the filter, both in the passband and in the *entire* stopbands.

We will limit our designs to five popular bandpass shapes—*maximum-peakedness, flattest-amplitude,* and shapes with 1-, 2-, or 3-dB passband dips.

SOME TERMS

Fig. 5-1 shows some typical bandpass filter shapes. The bandpass shape provides lots of attenuation to very low and very high frequencies, and much less attenuation to a band of medium frequencies. The actual values of the attenuation depend on the complexity (order) of the filter, the passband smoothness (relative Q or damping), and the gain or loss values designed into the filter.

A bandpass filter is used when we want to emphasize or pass a narrow signal band while attenuating or rejecting higher- or lower-frequency noise or interfering signals.

The *bandwidth* of the filter is defined as the difference between the upper and lower points where the filter response *finally* falls to 3 dB below its peak value on the way out of the passband. Our definition of bandwidth is *always* made 3 dB below peak, even if there is some other amount of passband ripple.

The *center frequency* of the filter is the *geometric mean* of the upper and lower 3-dB *cutoff frequencies* (Fig. 5-1). Sometimes the center frequency of a one-pole bandpass filter is called the *resonance*

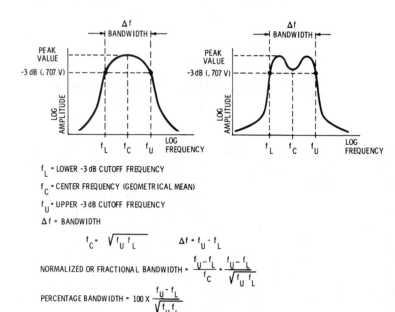

f_L = LOWER -3 dB CUTOFF FREQUENCY

f_C = CENTER FREQUENCY (GEOMETRICAL MEAN)

f_U = UPPER -3 dB CUTOFF FREQUENCY

Δf = BANDWIDTH

$$f_C = \sqrt{f_U f_L} \qquad \Delta f = f_U - f_L$$

$$\text{NORMALIZED OR FRACTIONAL BANDWIDTH} = \frac{f_U - f_L}{f_C} = \frac{f_U - f_L}{\sqrt{f_U f_L}}$$

$$\text{PERCENTAGE BANDWIDTH} = 100 \times \frac{f_U - f_L}{\sqrt{f_U f_L}}$$

Fig. 5-1. Bandpass filter shapes and terminology.

frequency. Note that the center frequency is *never* at "half the difference" between the upper and lower cutoff frequencies. It is *always* the square root of the product of the upper and lower cutoff frequencies.

Usually, we make the center frequency unity. Once the analysis and design are completed, the component values can be scaled as needed to get any desired center frequency.

The *fractional bandwidth* and the percentage *bandwidth* are two different ways of expressing the ratio of the bandwidth to the center frequency, with the formulas given in Fig. 5-1. The percentage bandwidth is always 100 times the fractional bandwidth. Either way is useful in suggesting an approach to a particular filter problem.

For instance, suppose we have a filter with an upper cutoff frequency of 1200 Hz and a lower cutoff frequency of 800 Hz. The center frequency will be 1000 Hz, right? *Wrong!* The center frequency is the geometric mean of the upper and lower cutoff frequencies, or 980 Hz. The bandwidth is simply the difference between the upper and lower cutoff frequencies, or 400 Hz. The fractional bandwidth will be the ratio between the bandwidth and center frequency, or 400/980, or 0.41. The percentage bandwidth is 100 times this, or 41%.

We can easily have a percentage bandwidth far in excess of 100%, although many filter problems usually deal with percentage bandwidths of under 50%. For instance, a bandpass filter to handle phone-quality audio from 300 Hz to 3000 Hz has a percentage bandwidth of 285%.

SELECTING A METHOD

The fractional bandwidth is the deciding factor in selecting the best filter for a particular filtering job. If the fractional bandwidth is very large, you do not build a "true" bandpass filter; instead, you get the passband by overlapping a high-pass and a low-pass filter. Our phone audio example covering 300 to 3000 Hz could best be done by using cascaded high-pass and low-pass sections. On the other hand, a modem data filter might cover 900 to 1300 Hz. It has a percentage bandwidth of only 37% and is best done by using the "true" bandpass techniques of this chapter and Chapter 7.

If the percentage bandwidth is less than 80 to 100%, use the "true" bandpass methods of Chapters 5 and 7.

If the percentage bandwidth is more than 80 to 100%, use an overlapping high-pass and low-pass filter cascaded, using the methods of Chapters 4, 6, and 8.

Fig. 5-2. Picking a bandpass design method.

You should usually make your decision around 80 to 100% bandwidth. Wider than this, and you will get better performance and a simpler design, with overlapping high-pass and low-pass sections, by following the methods of Chapters 4, 6, and 8. Narrower than this, and you should use the methods of this chapter and Chapter 7. Fig. 5-2 sums up this rule.

FILTER-SHAPE OPTIONS

One, two, three, or more cascaded *poles* can be used for the band-pass filter. Since each pole is a second-order active bandpass section, the order of the filter is twice the number of poles, resulting in second-, fourth-, or sixth-order filters.

With a single second-order bandpass section, all we can control is the center frequency and the damping, or its inverse, Q. The Q sets the bandwidth directly, as 1/Q will be the bandwidth of a unity-normalized filter.

With two cascaded sections, we can individually control the resonant frequency of each section and the Q of each section. For a balanced or symmetric response, it is best to keep both Qs identical. We can generate shape options by spreading the frequencies of the two sections symmetrically away from the center frequency.

For instance, if we place both sections at the same frequency, we get a very sharp response which we might call a *maximum-peakedness* response. The selectivity will be very high, but the passband will have a lot of droop.

Suppose we start spreading out the two poles in frequency while keeping their Q constant. This is done by introducing a factor "a" that is near unity. You *multiply* one frequency by "a" and *divide* the other one by "a," and the two poles split symmetrically from the center frequency.

As the spacing is increased, the passband becomes flatter and flatter, until a shape option called a *flattest-amplitude* filter is reached. More spacing, and we get a dip in the passband with 1, 2, or 3 dB of sag. All of these are useful shapes also. More spacing yet, and the filter breaks down into a useless two-humped response.

Insertion loss of the filter also goes up as the poles are spaced apart, but this is easily made up, as we will shortly see.

When three poles or three cascaded sections are used, we have many options, for each pole can have a specified frequency and Q, and we can spread them in different ways. You get useful results if you make two poles move and leave the third one at the center frequency. Again, a value "a" (different from the two-section case) is used to multiply one filter section and divide the other one in frequency. To keep the passband reasonably shaped, we set the Q of the outside two sections to some identical value and set the inside section to a Q value of *one-half* the outside two.

Once again, we generate different shape options as we vary "a," starting out with maximum-peakedness, flattest-amplitude, and on through the 1-, 2-, and 3-dB-dip filters. This time there are three peaks and two valleys in the response. The bandwidth is controlled by changing the Q of each section. Obviously, there is a unique rela-

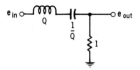

(A) Passive RLC.

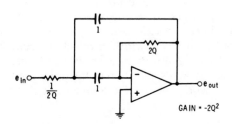

(B) Multiple feedback (moderate Q).

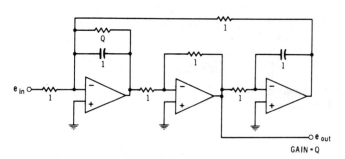

(C) Biquad (high Q).

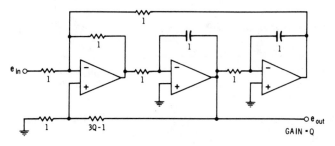

(D) State-variable (high Q).

Fig. 5-3. Single-pole or second-order active bandpass circuits.

tionship between Q and "a" that must be carefully controlled if we are going to get the shape we desire for various bandwidths.

These 11 shape options should cover the majority of bandpass filter applications. Higher-order shapes are easily worked out for specialized applications. We will find out as we go along that the shape options with dips in them tend to require higher-Q sections and need tighter tolerances on their design, and that their ringing, overshoot, and transient response get considerably worse. These disadvantages are traded off for a faster initial rate of falloff outside the passband.

The problem now is to come up with some response curves that show what we can expect from the one-, two-, and three-section cascaded sections. Then we can turn this information "inside out" so that we can start with a bandwidth or rejection need, find out how to fill the need with a given number of poles, and determine what the Q, center frequency, and separation of the poles are to be. Following this, we have to work out some tolerance criteria that tell us how accurate we must be.

SECOND-ORDER BANDPASS FILTER

Circuits that will give us a single response pole or a second-order bandpass response are shown in Fig. 5-3. Chapter 7 will show easy methods for actually building and tuning these circuits.

The math behind the single-pole response is shown in Fig. 5-4, and a plot of a 10% bandwidth (Q = 10) pole is shown in Fig. 5-5.

THE MATH BEHIND   **A single response pole.**

Use the series RLC circuit of Fig. 5-4A for analysis:

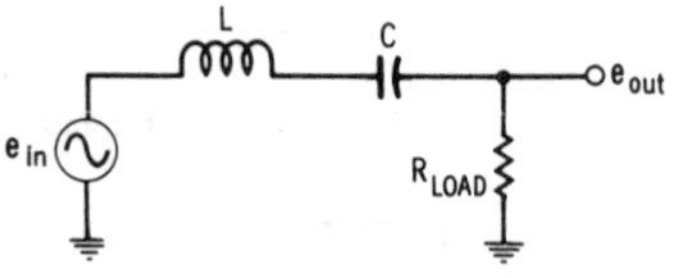

By voltage-divider analysis:

$$\frac{e_{out}}{e_{in}} = \frac{R_L}{R_L + j(X_L - X_C)} = \frac{R_L}{R_L + j\left(\omega L - \dfrac{1}{\omega C}\right)}$$

$$(j = \sqrt{-1}) \quad \textbf{(A)}$$

If we can get the results for any *one* normalized circuit, we can use the results for *all* possible circuits simply by scaling impedance and frequency. To do this, pick component values as follows:

Fig. 5-4.

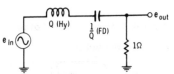

Note that with this choice of components $\omega = 2\pi f = 1$, since $\omega = \dfrac{1}{\sqrt{LC}}$ $= \dfrac{1}{\sqrt{Q/Q}} = 1$. The new output/input expression by substitution, is

$$\frac{e_{out}}{e_{in}} = \frac{1}{1 + i\left(\omega Q - \dfrac{Q}{\omega}\right)} = \frac{1}{1 + iQ\left(\omega - \dfrac{1}{\omega}\right)} = \frac{1}{1 + iQ\left[\dfrac{\omega^2 - 1}{\omega}\right]} \quad \text{(B)}$$

The loss of the circuit is 1/gain or

$$\text{loss} = \frac{e_{in}}{e_{out}} = 1 + iQ\left[\frac{\omega^2 - 1}{\omega}\right] \quad \text{(C)}$$

This is a vector quantity. The magnitude of the loss will be

$$\sqrt{(1)^2 + \left(Q\left[\frac{\omega^2 - 1}{\omega}\right]\right)^2} = \sqrt{1 + Q^2\left(\frac{\omega^2 - 1}{\omega}\right)^2} \quad \text{(D)}$$

Expressed in decibels, the loss is

$$\frac{e_{in}}{e_{out}} = 20 \log_{10} \sqrt{1 + Q^2\left(\frac{\omega^2 - 1}{\omega}\right)^2} \quad \text{(E)}$$

or the gain is

$$\frac{e_{out}}{e_{in}} = -20 \log_{10}\left[1 + Q^2\left(\frac{\omega^2 - 1}{\omega}\right)^2\right]^{1/2} \quad \text{(F)}$$

Note that the loss is minimum (0 dB) at $\omega = 1$ and increases for all other values of ω. The very same response will be obtained for any frequency of resonance by simple scaling. Thus

$$\boxed{\begin{array}{c} \dfrac{e_{out}}{e_{in}} = -20 \log_{10}\left[1 + Q^2\left(\dfrac{f^2 - 1}{f}\right)^2\right]^{1/2} \quad \text{(G)} \\[2mm] \text{single-pole response } f_C = 1 \end{array}}$$

The equivalent "S plane" expression for $\dfrac{e_{out}}{e_{in}}$ is found by letting $S = i\omega$ in (B) above. The result is

$$\frac{e_{out}}{e_{in}} = \left[\frac{1}{Q}\right]\left[\frac{S}{S^2 + \dfrac{1}{Q}S + 1}\right] \quad \text{(H)}$$

Fig. 5-4—continued.

To find a frequency for a specific attenuation, note that $\frac{f^2 - 1}{f} = \Delta f$ and rewrite (G). Let x = attenuation

$$x = \sqrt{1 + Q^2 \left(\frac{f^2 - 1}{f}\right)^2} = \sqrt{1 + Q^2 \Delta f^2}$$

$$x^2 = 1 + Q^2 \Delta f^2$$

$$\sqrt{x^2 - 1} = Q \Delta f$$

so,
$$\Delta f = \frac{\sqrt{x^2 - 1}}{Q} \tag{I}$$

For instance, Q = 30 and 20 dB attenuation, x = 10.0

$$\Delta f = \frac{\sqrt{100 - 1}}{30} = \frac{9.94}{30} = .331$$

To relate Δf, f_U and f_L we already have these expressions:

$$\Delta f = f_U - f_L \qquad f_C = \sqrt{f_U f_L}$$

During analysis, *always* let $f_C = 1$. So $\sqrt{f_U f_L} = 1$ and $f_U = \frac{1}{f_L}$

Now the normalized $\Delta f = f_U - f_L = f_U - \frac{1}{f_U}$

$$\Delta f = \frac{f_U^2 - 1}{f_U}$$

Rearranging

$$f_U^2 - \Delta f \, f_U - 1 = 0$$

This is a quadratic. Solving:

$$\boxed{\begin{array}{l} f_U = \dfrac{\Delta f + \sqrt{(\Delta f)^2 + 4}}{2} \text{ and } f_L = \dfrac{\Delta f - \sqrt{(\Delta f)^2 + 4}}{2} \\[2mm] f_U = 1/f_L \qquad\qquad\qquad f_L = 1/f_U \\[1mm] \text{Relating } f_U, f_L, \text{ and } \Delta f \text{ when } f_C = 1 \end{array}} \tag{J}$$

Fig. 5-4—continued.

There are several very important things to notice about Fig. 5-5. First, the response is symmetrical on a log frequency scale, with the same response at one-half and at two times the center frequency, at one-fourth and at four times the center frequency, and so on. To get

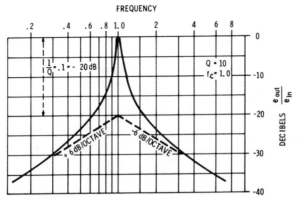

Fig. 5-5. Frequency responses of a single-pole, Q = 10 filter (see text).

symmetrical curves like these, you must use a log scale on your plot. Linear scales will skew or distort the picture.

The gain is highest at the center frequency. In a passive RLC series circuit with a good inductor, the input voltage will equal the output voltage at the resonance frequency, and the gain will be unity, or zero decibels. With an active second-order bandpass, you will see that you can make the gain anything you like, although it often turns out to be related to the bandwidth and Q for a specific circuit.

As we go up or down in frequency from resonance, more and more insertion loss is picked up. The −3-dB points (0.707 voltage ratio) with respect to the center-frequency value are called the *cutoff frequencies*, while the difference between the upper and the lower cutoff frequencies is called the *bandwidth*.

We call the inverse of the bandwidth Q. A Q of 85 is a fractional bandwidth of .012, or a percentage bandwidth of 1.2%. Note that Q is the inverse of the −3-dB bandwidth *only* if a single section is in use.

As we go farther from the center frequency, the response starts dropping off very sharply, expressed as so many decibels per octave of *bandwidth*. But, *sooner or later the curves start flattening out,* giving much less rejection than you might except from looking at the initial rate of falloff. Why?

The answer is easiest to see in the case of the passive RLC circuit. Near resonance, all three components are active in the response. But, at very high frequencies the reactance of the capacitor is negligible, and the circuit degenerates to a simple series-RL circuit dropping at a 6-dB-per-octave absolute frequency rate.

Similarly, at very low frequencies the reactance of the inductor becomes very small, and the circuit looks like a series-RC circuit, which also drops at a 6-dB-per-octave absolute frequency rate in the opposite direction. *Regardless* of how high the Q is or how fast the response drops off initially, the *ultimate* response falloff is only 6 dB per octave.

We can easily generate a family of curves or a listing by using the math of Fig. 5-4. Fig. 5-6 is a plot of the Q versus the 1-, 3-, 10-, 20-, 30-, and 40-dB bandwidths, the upper-frequency cutoff points, and the lower-frequency cutoff points. This plot contains all that is needed to completely specify a single-pole bandpass filter to nominal accuracy. Fig. 5-7 gives some examples of how to use this response curve.

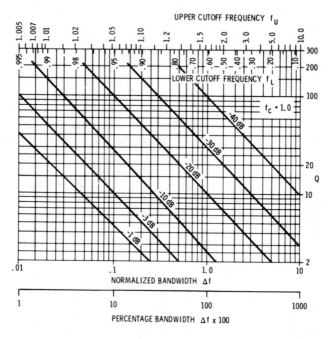

Fig. 5-6. Complete response characteristics of a single-pole bandpass filter.

TWO-POLE, FOURTH-ORDER BANDPASS RESPONSE

Single-pole filters usually do not offer enough stopband attenuation to be very useful, particularly at frequencies far into the stopband. Their performance can be dramatically improved by adding a second section, particularly if it is staggered slightly in frequency from the first one.

Single-pole bandpass filter design

1. Sketch the response plot of a Q = 60, single-pole bandpass filter centered at 1 kHz without using any math.

   The maximum response will be zero decibels at 1 kHz. From Fig. 5-6, the 3-dB bandwidth is read as a normalized .0165. Multiply this by 1000 to get the bandwidth in hertz, or 16.5 Hz to the —3-dB points. The upper cutoff frequency will be 1.0085 times the center frequency, or 1.0085 Hz; the lower cutoff frequency will be .9915 times the center frequency, or 991.5 Hz. We then read the —10-dB points as 975 Hz and 1025 Hz. The —20-dB points are 918 Hz and 1089 Hz. The —30-dB points are 770 Hz and 1290 Hz and the —40-dB points are 480 Hz and 2100 Hz. The plot can now be accurately sketched on semilog paper.

2. Sketch the response plot of a Q = 60, single-pole bandpass filter centered at 250 Hz.

   We simply use the results of example No. 1 and move everything to the new frequency by multiplying all the answers by 250/1000, the frequency-scaling factor. Results are as follows:

Center frequency	250
Bandwidth	4.125 Hz
—3 dB	247.8 & 251.9 Hz
—10 dB	243.7 & 256.25 Hz
—20 dB	229.5 & 272.25 Hz
—30 dB	192.5 & 322.5 Hz
—40 dB	120 & 525 Hz.

3. What is the maximum —3-dB bandwidth of a 2-kHz single-pole bandpass filter if it is to give at least 30 dB of rejection to a 1500-Hz interfering signal?

   The 1500 Hz is 1500/2000 times the center frequency, or an f_L of 0.75. From Fig. 5-6 at 0.75 f_L and 30-dB loss, we read a Q of 56, and a corresponding normalized bandwidth of .0175. When the normalized —3-dB bandwidth of .0175 is multiplied by the 2000-Hz center frequency, a final bandwidth of 35 Hz results.

   Note that this example simply calls for 30 dB of attenuation to an interfering signal. This in NO WAY guarantees that all interfering signals will be 30 dB below the desirable ones; all it means is that any interfering signal comes *out* of the filter 30 dB lower than it went in. For instance, if your interference starts out 40 dB stronger than the signal you want, the filter will provide the 30-dB rejection you asked it to, only you will still end up with a —10-dB signal-to-noise ratio. When you establish how much rejection is to be provided, the relative signal levels and dynamic ranges must be taken into account.

Fig. 5-7.

For a symmetrical response, make the Qs of both poles identical. The ratio of the pole Q to the staggering will set the shape of both the passband and the stopband. Outside of this new staggering factor, the math and the design are just about the same as for the single-pole filter.

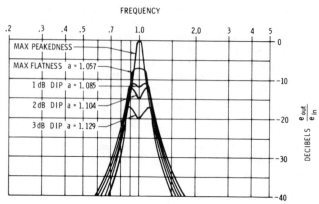

Fig. 5-8. Frequency response of a two-staggered-pole, Q = 10 filter (see text and Fig. 5-12).

In Fig. 5-8, we have plotted several Q = 10 pole pairs for different amounts of staggering, while Fig. 5-9 shows the related mathematics.

The staggering factor is called "a." Usually, "a" is a number slightly more than unity. It *multiplies* the center frequency of the upper-pole location and *divides* the center frequency of the lower-pole location.

THE MATH BEHIND   **Two-pole bandpass response.**

From Fig. 5-4(G), the single-pole filter response is

$$\frac{e_{out}}{e_{in}} = 20 \log_{10}\left[1 + Q^2\left(\frac{f^2 - 1}{f}\right)^2 \right]^{1/2}$$

$$(f_C = 1) \quad (A)$$

Cascading two of these will give us their product, or

$$\frac{e_{out}}{e_{in}} = 20 \log_{10}\left[\left(1 + Q^2\left\{\frac{f^2 - 1}{f}\right\}^2\right)^{1/2} \left(1 + Q^2\left\{\frac{f^2 - 1}{f}\right\}\right)^{1/2} \right] \quad (B)$$

This expression directly handles the maximum-peakedness case. For pole pairs symmetrically different from $f_C = 1$, a term "a" is introduced that is multiplied by the center frequency of one pole and divided by the other. The expression becomes

Fig. 5-9.

$$\frac{e_{out}}{e_{in}} = 20 \log_{10}\left[\left(1 + Q^2\left\{\frac{f^2a^2 - 1}{fa}\right\}^2\right)^{1/2}\right.$$

$$\left.\left(1 + Q^2\left\{\frac{f^2/a^2 - 1}{f/a}\right\}^2\right)^{1/2}\right] \qquad \text{(C)}$$

amplitude response, two poles staggered by a, $f_C = 1$

The problem now consists of finding a, Q, and f values for useful filter shapes. Since there is no sane way of solving (C) for a, Q, or f, trial-and-error substitution is the best and easiest way, easily done on a hand calculator, programmable calculator, or BASIC computer program.

If $Q = 10$ is picked for analysis, the "a" values for maximum peakedness, maximum flatness, and 1-, 2-, and 3-dB dips are 1.000, 1.057, 1.085, 1.104, and 1.129, respectively. Once Q and a values are found, the $Q = 10$ curves can be plotted.

Insertion loss is given by solving at $f_C = 1$ and is

$$\frac{e_{out}}{e_{in}} = 20 \log_{10}\left[1 + Q^2\left(\frac{a^2 - 1}{a}\right)^2\right] + K \qquad \text{(D)}$$

insertion loss, two-pole staggered filter

K is 0 for maximum peakedness and maximally flat responses. It is $+1$ dB for a 1-dB dip, $+2$ dB for a 2-dB dip and $+3$ dB for a 3-dB dip. You always measure the insertion loss to the peak value, even when it does not occur at $f_C = 1$. For instance, insertion loss of a 2-dB dip filter is its loss calculated at $f_C = 1$ or -13.2 dB *plus* its peak value ($+2$dB) or -11.2 dB.

For different bandwidths, new pairs of Q and a have to be calculated. All these curves will have identical insertion loss at $f_C = 1$ for a given selected shape. Letting $f = 1$ in (C) and requiring a constant insertion loss tells us that

$$1 + Q^2\left(\frac{a^2 - 1}{a}\right)^2 \text{ has to be a constant.}$$

which means that $Z = Q\left(\frac{a^2 - 1}{a}\right)$ must be constant for a given response shape. To find a new a, Q pair, let

$$Z = Q\frac{a^2 - 1}{a} \text{ or}$$

Fig. 5-9—continued.

103

$$a^2 - \frac{Z}{Q}a - 1 = 0.$$ Once again, this is a

quadratic and solving for the larger value of a gives us

$$a = \frac{\dfrac{Z}{Q} \pm \sqrt{\left(\dfrac{Z}{Q}\right)^2 + 4}}{2} \qquad \text{(E)}$$

a for a given Q and given shape.

Once all the a, Q pairs are found, response can be plotted for various bandwidths. If we find the −3-, −10-, −20-, −30-, −40-dB points for the Q = 10 pairs, we can predict values for other Qs by letting QΔf = constant, and solving for new fs corresponding to a new Q value.

Values of extremely low Q (less than Q = 3) have to be separately calculated. Insertion losses for ultra-low Q are also higher than (D). Qs less than 2 are not normally useful as filters.

Fig. 5-9—continued.

As "a" increases, the passband shape first gets much flatter; then it picks up a progressively deeper ripple trough, at the same time giving a sharper initial falloff outside the passband. Larger values of "a" break the response into a two-humped one that usually is not useful.

As "a" gets bigger, an insertion loss appears, which turns out to be constant for any particular response shape, regardless of the pole Q and the bandwidth. This loss is independent of frequency and easily made up elsewhere in the circuit. Many active bandpass circuits provide a lot of gain which is usually more than enough to offset this insertion loss and then some, particularly with higher Qs and narrower bandwidths.

Fig. 5-10 shows the insertion losses for the various shape options. These range from zero dB for the maximum-peakedness option up through 17 dB for the −3-dB-trough shape. If we subtract these fixed insertion losses from the curves of Fig. 5-8, the new plot of Fig. 5-11 results. This makes the benefits of passband dips more obvious. Note that the insertion loss is always defined with reference to the peak of the response curve and the bandwidth is defined with reference to 3 dB below the peak on the way out of the passband.

Maximum peakedness	0 dB
Maximum flatness	7.0 dB
1-dB dip	11.2 dB
2-dB dip	14.0 dB
3-dB dip	17.0 dB

Loss is measured from the *peak* of the response. Center frequency attenuation of 1-, 2-, and 3-dB dip filters will *exceed* their respective losses by 1, 2, and 3 dB.

Fig. 5-10. Insertion losses of two-pole bandpass filters (Q ≥ 3).

The Q, the staggering, the "a" value, and the bandwidth no longer have a simple relationship as was the case for a single-pole filter. *A specific Q value must be paired with a specific "a" value for a given response shape.* This is plotted in Fig. 5-12.

Bandwidth-versus-Q plots appear in Figs. 5-13 through 5-17 for the five shape options. Note that the selection of a shape option and a Q value uniquely defines the needed "a" value, shown in Fig. 5-12. An example showing how to use the two-pole curves appears in Fig. 5-18.

The maximum-peakedness curve is seldom used because of the steepness of its passband and its inefficiency. The best overall choice is usually the flattest-amplitude curve, except where "every inch" of initial stopband rejection is needed. Also, the flattest-amplitude response gives better transient and overshoot performance, takes lower-Q poles of looser tolerance, and has less insertion loss.

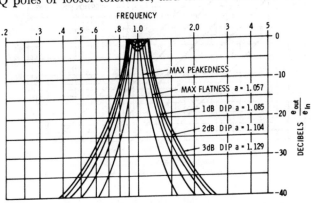

Fig. 5-11. Q = 10, 2-pole response redrawn from Fig. 5-8 with insertion losses removed. Note the steeper falloff of curves with dips in them. Accurate values can be read from Figs. 5-13 through 5-17.

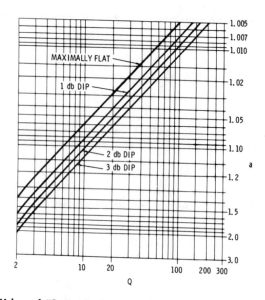

Fig. 5-12. Values of "Q-a" pairs for a given-response-shape, 2-pole bandpass filter.

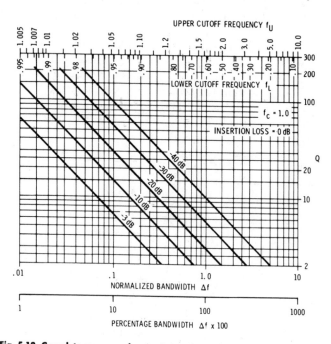

Fig. 5-13. Complete response characteristics of a two-pole, maximum-peakedness bandpass filter.

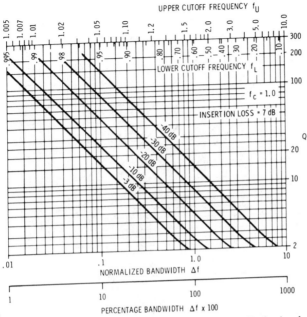

Fig. 5-14. Complete response characteristics of a two-pole, maximally flat bandpass filter.

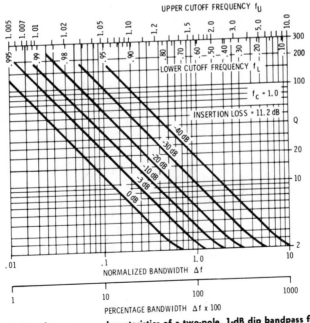

Fig. 5-15. Complete response characteristics of a two-pole, 1-dB dip bandpass filter.

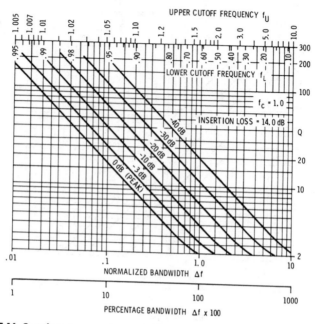

Fig. 5-16. Complete response characteristics of a two-pole, 2-dB dip bandpass filter.

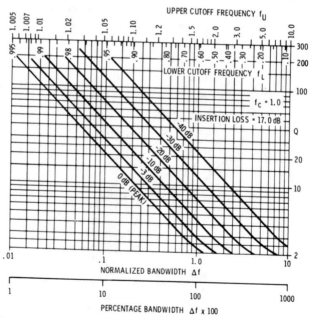

Fig. 5-17. Complete response characteristics of a two-pole, 3-dB dip bandpass filter.

Using the two-pole curves; an EXAMPLE

PROBLEM: An octave-wide two-pole filter is to cover 200 Hz to 400 Hz with a 1-dB dip in the passband.

(a) What will the rejection be to 50, 100, 800, and 1600 Hz?

(b) What are the Q values and pole frequencies for the two poles?

SOLUTION: The filter has a bandwidth of 200 Hz and a center frequency of $\sqrt{200 \times 400} = 283$ Hz. The bandwidth is $200/283 = .706$ or 70.6%.

(a) 100 Hz will be at a frequency of 100/283 or .3533 times the center frequency. From Fig. 5-15 we see that a bandwidth of .706 to the 3-dB points corresponds to a Q of 3.2. The attenuation of a $Q = 3.2$ pole at .3533 times the center frequency is around 25 dB.

50 Hz will be at a frequency of 50/283 or .0176 times the center. Attenuation here for a Q of 3.2 will be about 40 dB. Similarly, 800 Hz will be at a frequency of 800/283 or 2.82 times the center frequency and will also have an attenuation of 25 dB. Finally, 1600 Hz will be 1600/283 or 5.653 times the center frequency and will also be attenuated 40 dB.

(b) From part (a) we know that the Q of each pole is to have a value of 3.2. This Q must be associated with a unique "a" value to get the 1-dB dip shape. From Fig. 5-12, we read "a" as 1.32.

The upper pole location will be at 283×1.32 or 374 Hz. The lower pole location will be at $283/1.32 = 214$ Hz.

Insertion loss will be 11.2 dB. If each of our active poles has a gain of "Q" (see Chapter 7), their cascaded gains will be $3.2 \times 3.2 = 10.24$, or 20.2 dB. Thus the *net* gain of the active circuit will be 9 dB, or 2.8 times the input voltage.

Fig. 5-18.

THREE-POLE, SIXTH-ORDER BANDPASS RESPONSE

The analysis of the three-pole filter is similar to that of the two-pole filter. We limit our choice of shapes to one that has two poles staggered from the center frequency by a factor "a," just as in the two-pole case, only "a" will have a slightly different value for a given response.

Next, we add a new pole at the center frequency. We set the Q of this new pole, which is not moved by "a," to *one-half* the Q of the two outer poles, so that its contribution to the bandwidth is essentially *twice* that of the others.

With the three-pole response, a maximum always occurs at the center frequency. If "a" is unity, we have the maximum-peakedness case. Fig. 5-19 shows the effects of spacing the two outer poles by

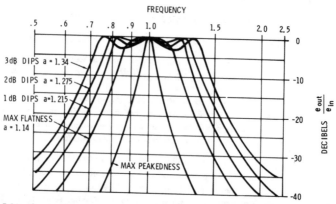

Fig. 5-19. Three-pole response; outer poles, Q = 10; center pole, Q = 5. Note change of frequency scale from Fig. 5-11. These curves are much steeper for a given bandwidth.

"a" for a Q = 10 value case. The center pole is held to a Q = 5. As with Fig. 5-11, the insertion loss has been removed from all the curves so that their shapes can be better compared. The curves with dips in them have a maximum value at the center frequency, a dip, and then a second pair of maximum values above and below the dip frequencies, giving three humps to the curves. If for the center pole we pick some value of Q that is not one-half the value of the outer ones (actually very slightly less than one-half for the dip cases, but not worth bothering over), the outer peak amplitudes will be above or below the middle one, which gives us a "camel" response shape of limited usefulness.

The math behind the three-pole bandpass filter is shown in Fig. 5-20, and the fixed insertion losses we can expect appear as Fig. 5-21. Values for very low Q (less than three) will be somewhat higher than these figures. The relationship between (Q) and "a," which is needed for certain shapes, is given in Fig. 5-22, while the complete response characteristics for the various shapes are shown in Figs. 5-23 through 5-27. Finally, Fig. 5-28 shows how to use the curves in an example.

COMPONENT TOLERANCES AND SENSITIVITIES

How accurate do the components and active circuits have to be for a given response shape? As a general rule, the narrower the bandwidth or the deeper the passband dips, the more accurate the com-

THE MATH BEHIND **Three-pole bandpass response.**

There are many possible Q and "a" arrangements. Nearly optimum results occur if a new pole of *one-half* the Q is added to the center ($f_C = 1$) of a two-pole response. Adding the new pole to the expression of Fig. 5-9C gives us

$$\frac{e_{out}}{e_{in}} = 20 \log_{10}\left[\left\{1 + Q^2\left(\frac{f^2a^2 - 1}{f^2a^2}\right)\right\}^{1/2}\right.$$
$$\left.\left\{1 + Q^2\left(\frac{f^2/a^2 - 1}{f/a}\right)\right\}^{1/2}\left\{1 + \left(\frac{Q}{2}\right)^2\left(\frac{f^2 - 1}{f}\right)\right\}^{1/2}\right] \quad \text{(A)}$$

three-pole response, two poles are Q-shifted by "a," one pole of *one-half* Q stationary at $f_C = 1$

Once again, trial and error is needed to find optimum "Q-a" pairs. For Q = 10, max peakedness "a" = 1.00; max flatness "a" = 1.14; 1-dB dips, a = 1.215; 2-dB dips, "a" = 1.275; and 3-dB dips, "a" = 1.34; with a center-pole Q = 5.
Since one peak value occurs at $f_C = 1$, the insertion loss remains as

$$\frac{e_{out}}{e_{in}} = 20 \log_{10}\left[1 + Q^2\left(\frac{a^2 - 1}{a}\right)^2\right] \quad \text{(B)}$$

three-pole insertion loss

Other "Q-a" pairs are predicted as in Fig. 5-12. Low-Q (less than 3) "a" values have to be separately calculated. Insertion losses for these values are also higher than Fig. 5-21 values. Q values less than 3 are rarely useful as 3-pole filters.

Fig. 5-20.

Maximum peakedness	0 dB
Maximum flatness	18.0 dB
1-dB dips	24.2 dB
2-dB dips	28.0 dB
3-dB dips	31.2 dB

Fig. 5-21. Insertion losses of three-pole bandpass filters ($Q \geq 3$).

ponent choices have to be. Except for very wide bandwidth filters, some means of adjusting or tuning the center frequency should be provided. For very narrow filters, a means of adjusting Q is handy, as well, although the frequency adjustment is often far more important.

Calculating real sensitivities is a painful process; you tend to lose insight into what is really happening when you get into the math details. Instead of this route, you can directly estimate the sensitivities by using Figs. 5-6, 5-12, and 5-22, as shown in the example of Fig. 5-29.

Generally, you estimate how much f, Q, or "a" can be shifted before the shape of the final response curve is altered enough that you do not want it or cannot use it. This gives you an upper bound on the allowable component and circuit variations.

USING THIS CHAPTER

The curves of this chapter tell you exactly what the response shape of a given type of bandpass filter will be for all frequencies, both near and far from the passband. The same curves do the essential conversion from a filter need to a set of values that can be cranked into the circuitry of Chapter 7. Finally, these curves tell you fairly accurately just how precise your need is and what range of adjustment you should include on your filter.

The following is a set of rules for use of this chapter:

1. Specify what you want to pass and what you want to block in terms of bandwidths, center frequencies, and rejection frequencies. Normalize this to a unity center frequency.
2. If the bandwidth is over 80 to 100%, use cascaded high-pass and low-pass sections. If it is lower, use the methods of this chapter.
3. From the curves, select the simplest, most heavily damped response characteristic that does the job. If three poles do not seem to do the job, chances are you have overspecified requirements. (For cases where you must have higher performance, consider a higher order, the elliptical filters of Chapter 9, or alternatives such as digital filters, phase-locked loops, quartz resonators, sideband techniques, etc.)
4. From your choice of shape, specify the Q, "a," and frequency values needed and convert them to actual frequencies in order to spell out the real-world parameters you will need for your filter.
5. Estimate the component tolerances and sensitivity using the guidelines of Fig. 5-29.
6. Using the Q, f, "a," and insertion-loss values, go to Chapter 7 and build the filter.

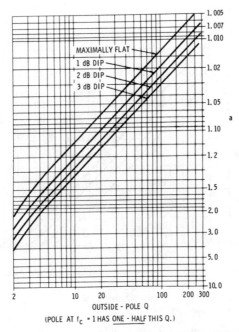

Fig. 5-22. Values of "Q-a" pairs for a given-response-shape, 3-pole bandpass filter.

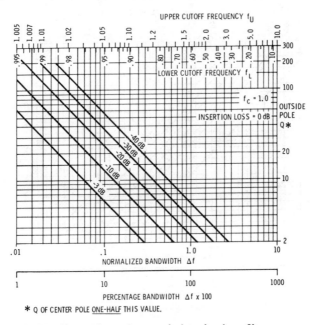

Fig. 5-23. Three-pole, maximum peakedness bandpass filter response.

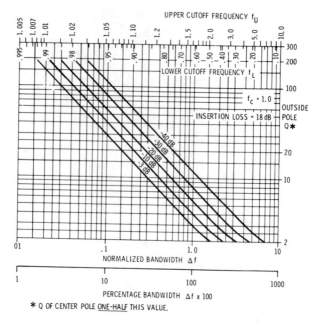

Fig. 5-24. Three-pole, maximally flat bandpass filter response.

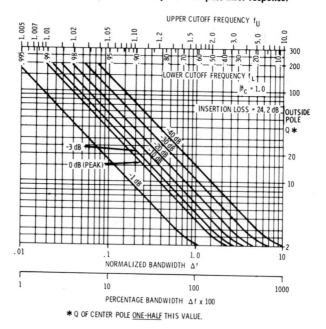

Fig. 5-25. Complete response characteristics of a three-pole, 1-dB dips bandpass filter.

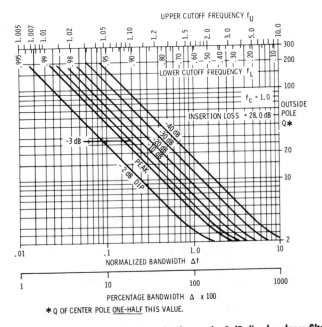

Fig. 5-26. Complete response characteristics of a three-pole, 2-dB dips bandpass filter.

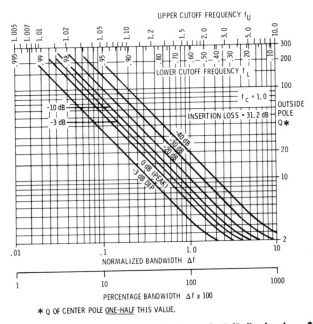

Fig. 5-27. Complete response characteristics of a three-pole, 3-dB dips bandpass filter.

Using the three-pole curves; an EXAMPLE

PROBLEM: A bandpass filter to be used in some brain-wave research is to have a good transient response, a passband from 7.0 to 8.0 Hz, and at least 40 dB of rejection to a frequency of 5 Hz. Specify the filter.

SOLUTION: The center frequency is $\sqrt{7 \times 8} = 7.48$ Hz. The bandwidth is $1/7.48 = .133$ or 13.3%. The 5-Hz reject frequency is $5/7.48 = .67$ when normalized to the center frequency.

We first try a single-pole filter, Fig. 5-7. For a 13.3% bandwidth, the rejection at .67 is only 17 dB or so, much too little. We then try the two-pole filters of Figs. 5-14 through 5-18. Even with the 3-dB dip, the attenuation at .67 frequency is only 38 dB. We might be tempted to cheat a bit and use the simpler circuit, except for the fact that we also need good transient performance, and this is the most poorly damped of all the two-pole filters. So, we try the three-pole curves.

Fig. 5-24, a maximum-peakedness filter, cannot do the job, so we try the maximally flat curve of Fig. 5-25, which easily gives us well over the needed 40 dB. The Q value of 22 is used directly for the outer poles and a Q of 11 is used at the center pole. From Fig. 5-23, the corresponding "a" value is 1.06. This is multiplied by one pole and divided by a second and ignored by the third. The final pole frequencies and Qs would be

---

Lower: Q = 22

f = 7.48/1.06 = 7.05 Hz

Center: Q = 11

f = 7.48 Hz

Upper: Q = 22

f = 7.48 × 1.06 = 7.92 Hz

The tolerances could be estimated as in the next example. While any of the three-pole curves with deeper passband dips would also work and even give us more attenuation, a price in tolerance and ringing would have to be paid. The maximally flat filter is usually the best overall choice. If the n = 3, 3-dB dip filter or less cannot do a defined filter job, then the specs are probably too tight to be done reasonably with *any* active filter technique.

Fig. 5-28.

Frequency and Q tolerance and sensitivity; an   **EXAMPLE**

PROBLEM: A certain two-pole, 1-dB dip, bandpass filter needs pole Q values of 10 and "a" values of 1.09. How accurate do we have to be in the final circuit?

SOLUTION: We estimate accuracy by setting up some limit of degradation of response that is acceptable. With 1-dB dip filter, letting it become a 2-dB dip one sets one possible response limit. We can directly estimate frequency accuracy by shifting the "a" value on Fig. 5-12, and we can estimate the Q accuracy by shifting Q to these limits.

Frequency: The nominal "a" value of 1.09 increases to 1.11 for a 2-dB dip at Q = 10. This is a change of (1.11 — 1.09)/1.09 = .0183, or 1.83%. Thus a frequency tolerance somewhat tighter than 2% is needed.

Q Accuracy: The nominal "Q" value of 10 increases to 12 for a 2-dB dip at "a" = 1.09. This is a change of (12 — 10)/10 = .2, or 20%. Q may change 20% to reach this limit.

Generally, the Q tolerance is less critical than the frequency tolerance. Tighter restrictions are associated with higher Qs, narrower bandwidths, deeper dips, and higher order responses.

Fig. 5-29.

CHAPTER 6

Low-Pass Filter Circuits

Chapter 4 showed how to take a need for a low-pass filter and convert this need into a specification for a filter of a certain order and shape option. This information was then converted by the tables of Chapter 4 into listings for a group of cascaded first- and second-order sections. The frequency of the first-order sections and the frequency and damping values of the second-order sections were combined to get the desired overall shape.

In this chapter, we will consider how to actually build active first- and second-order circuits. We will learn how to cascade them properly to get what we are after. For those who are interested in the "why" of these circuits, we will give a detailed analysis of what the circuits do and what options they can give you. As usual, we will avoid any advanced math.

Finally, we will present a catalog of instant design, math-free, "ripoff," low-pass-filter, *final circuits* that are ready for instant use, and we will give a few examples.

TYPES OF LOW-PASS FILTERS

There are two basic types of low-pass filters. One is the "true" low-pass filter; the other is the "ac-coupled" low-pass filter. As Fig. 6-1 shows us, there are very specific restrictions that influence your choice of which one to use.

The response of a true low-pass filter extends down to dc, which means that any input voltage levels, bias shifts, etc., are accepted and passed on to the output of the filter. If the input voltages are too large, they will restrict the dynamic range of the filter or drive it into clipping, limiting, or saturation. One advantage of true low-pass

1. The response of a true low-pass filter extends down in frequency to dc. Any voltages, bias levels, drifts, offsets, etc. at the input get passed on to the output. Such dc "signals" can saturate the filter or otherwise restrict its dynamic range.
2. If a low-pass filter has its input "ac coupled" to block dc input voltages and offsets, some *internal* dc bias path to ground MUST be provided for proper filter operation.

Fig. 6-1. Input restrictions on an active low-pass filter.

filtering is that there are no transient effects caused by coupling capacitors when inputs are suddenly shifted or changed. Furthermore, the input signals are continually referenced to a specific base line or voltage level.

In an ac-coupled low-pass filter, we simply put a blocking capacitor on the input to pass the signal but block any bias levels or dc offsets on the input. Actually, what we have done is converted the filter into a bandpass filter whose lower cutoff frequency is sloppily determined by the time constant of the input capacitor and whose upper cutoff frequency is precisely set by the active low-pass filter.

For most audio filtering applications, ac coupling, with the lower frequency being set at a few hertz, is best, even though thumping, switch transients, and lack of dc restoration can be introduced. There is, however, one major and serious restriction to all ac-coupled low-pass filters:

On ALL active low-pass filters, some dc bias path to ground MUST be provided at all times for proper biasing of the operational amplifiers used in the filter.

If we ac couple, this means we must provide this bias return path internally to the filter. Fig. 6-2 shows how we can use an operational-amplifier voltage follower both to ac couple the input and to provide a ground reference dc bias return for the active filter stages that follow. Gain or loss can also be added to this stage to adjust system levels, if this is needed.

FIRST-ORDER LOW-PASS CIRCUITS

There is only one basic first-order, low-pass, active section, as shown in Fig. 6-3. It consists simply of a passive RC low-pass filter with an op-amp voltage follower on the output. The voltage follower

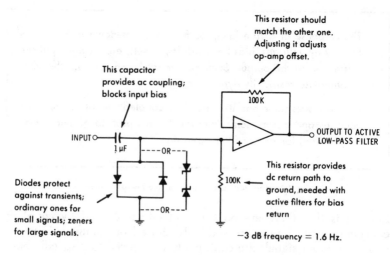

This resistor should match the other one. Adjusting it adjusts op-amp offset.

This capacitor provides ac coupling; blocks input bias

100 K

INPUT

1 µF

--- OR ---

--- OR ---

Diodes protect against transients; ordinary ones for small signals; zeners for large signals.

100K

OUTPUT TO ACTIVE LOW-PASS FILTER

This resistor provides dc return path to ground, needed with active filters for bias return

−3 dB frequency = 1.6 Hz.

Fig. 6-2. One method of ac coupling a low-pass filter.

isolates any output loads and prevents them from loading the capacitor. Normally, the gain is precisely unity, but gain can be added by providing a second resistor from the inverting input to ground. As with all active low-pass filters, the input must be provided with a low-impedance, dc bias return path to ground.

We will show two normalized versions of many of the circuits we will be using. One is convenient for analysis and is based on a 1-ohm and 1-radian-per-second reference. The second version is more convenient for actual use and is centered on a 10K impedance level and a 1-kHz cutoff frequency. To shift the frequency from 1 kHz, just read a new capacitor value from Fig. 6-21 or else just change the capacitor inversely with frequency. Doubling the capacitance halves the frequency, and so on.

The feedback resistor from output to inverting input is not critical in value, and in simple applications may be replaced with a short.

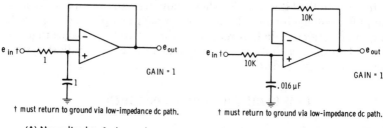

e_{in} †

1

1

e_{out}

GAIN = 1

† must return to ground via low-impedance dc path.

(A) Normalized to 1 ohm and 1 radian/sec.

e_{in} †

10K

10K

.016 µF

e_{out}

GAIN = 1

† must return to ground via low-impedance dc path.

(B) Normalized to 10K and 1-kHz cutoff frequency.

Fig. 6-3. First-order low-pass active section.

Its optimum value should match the resistance on the noninverting input to minimize op-amp offset. In critical applications, you can adjust this resistor to minimize the offset.

For very low frequencies, the impedance level of this circuit can be raised to 100K, which multiplies the resistors by ten and divides the capacitors by ten. Capacitors will be much smaller, but op-amp offset will be more of a problem.

The cutoff frequency of the first-order low-pass circuit can be tuned by switching the capacitor or by varying the resistor over a reasonable range. For instance, a 10:1 resistance variation will give a 10:1 cutoff-frequency variation, with the *higher* frequency associated with the *lower* resistance value.

Theoretically, first- and second-order sections can be cascaded in any order, but to prevent very large out-of-band signals from ringing or clipping, it always pays to put the first-order section *first* in any cascading, and second-order sections should follow with the highest-damped (large d values) sections towards the input and the lowest-damped (smallest d values) sections towards the output. In this way, large unwanted signals are cut down considerably before they get to a stage where they could ring, clip, or introduce other distortion.

SECOND-ORDER LOW-PASS CIRCUITS

There are many possible second-order active low-pass circuits, even if we eliminate the ones that are difficult or impossible to tune, that have too high a component sensitivity or output impedance, that interact, or that are hard to design or otherwise impractical. My choice in this text is to use four basic second-order active sections with various available features. Two of these are called *Sallen-Key* or VCVS (voltage-controlled voltage source) filters; two are called *state-variable* circuits. Let's take a closer look.

UNITY-GAIN SALLEN-KEY CIRCUITS

This is the simplest second-order active filter you can build; we used it in Fig. 1-2D to show what an active filter had to offer. Second-order Sallen-Key circuits in general consist of two cascaded RC sections driving a high-input-impedance noninverting amplifier. Feedback from the output to one of the resistors or capacitors bolsters what would normally be a drooping, highly damped, cascaded RC response. This positive feedback provides enough extra gain near the cutoff frequency to give us any value of damping we might like. The op amp serves to route energy from the supply into the proper point of the circuit to provide a response with a mathematical equivalent identical to that of a single inductor-capacitor section—minus, of

Sallen-Key low-pass second-order sections.

Sallen-key second-order low-pass filters can usually be redrawn into a passive network with an active source that looks like this:

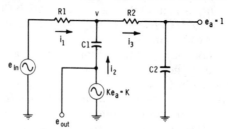

Since this network has to behave identically for any reasonable voltage at any point, it is convenient to let $e_a = 1$ volt and $e_{out} = Ke_a = K$. Solve for i_1, i_2, and i_3, then sum them:

$$i_3 = \frac{1 \text{ volt}}{Z_{C2}} = \frac{1}{\dfrac{1}{j\omega C2}} = j\omega C2$$

$$v = 1 + R2i_3 = 1 + j\omega C2 R2$$

$$i_1 = \frac{e_{in} - v}{R1}$$

$$i_2 = (K - v)j\omega C1$$

$$i_1 + i_2 = i_3$$

$$\frac{e_{in} - v}{R1} + (K - v)j\omega C1 = j\omega C2$$

$$e_{in} = j\omega R1 C1 + v - (K - v)j\omega R1 C1$$

$$e_{in} = (j\omega)^2 R1 C1 R2 C2 + (j\omega)[R1 C1 + C2 R2 + (1 - K)R1 C1] + 1$$

Letting $S = j\omega$ and dividing by $R1 C1 R2 C2$

$$e_{in} = R1 C1 R2 C2 \left[S^2 + \left\{ \frac{1}{R1 C1} + \frac{1}{R2 C1} + (1 - K)\frac{1}{R2 C2} \right\} S + 1/R1 C1 R2 C2 \right]$$

$$e_{out} = K$$

$$\frac{e_{out}}{e_{in}} = \frac{K/R1 C1 R2 C2}{S^2 + \left[\dfrac{1}{R2 C1} + \dfrac{1}{R1 C1} + (1 - K)\dfrac{1}{R2 C2} \right]S + 1/R1 C1 R2 C2}$$

Fig. 6-4.

For a useful filter, we would like to have frequency, gain, and damping independently adjustable, which places very definite restrictions on values of K and the ratios of the resistors and capacitors. If we can get the expression into a form of

$$\frac{e_{out}}{e_{in}} = \frac{K}{S^2 + dS + 1}$$

we have a second-order low-pass section. For frequencies *not* equal to unity, we must make the expression look like

$$\frac{e_{out}}{e_{in}} = \frac{K\omega^2}{S^2 + d\omega S + \omega^2} \tag{A}$$

Otherwise any attempt at changing damping will change gain or frequency, and vice versa. Getting things into this final form places very specific restrictions on gain and component values. Let us derive two possible useful, noninteracting circuits:

(1) *Unity gain, equal resistors:* Let R1 = R2 and K = 1. R1R2C1C2 = 1

for $\omega = 1$, so $C1 = \dfrac{1}{C2}$

$$\left[\frac{1}{R2C1} + \frac{1}{R1C1} + (1-K)\frac{1}{R2C2}\right]S = \left[\frac{1}{C1} + \frac{1}{C1}\right]S = \left[\frac{2}{C1}\right]S$$

$$\therefore d = \frac{2}{C1} \text{ and } C1 = \frac{2}{d} \text{ and } C2 = \frac{d}{2}$$

Further, for R1 = R2 = 1 we have expression (A) above, guaranteeing noninteracting adjustments. The circuit looks like this:

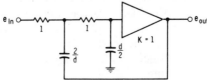

The amplifier can be an emitter follower or an op-amp voltage follower

(2) *All components identical:*

Let R1 = R2 = C1 = C2. For $\omega = 1$, R1 = R2 = C1 = C2 = 1

$$\left[\frac{1}{R2C1} + \frac{1}{R1C1} + (1-K)\frac{R2C2}{1}\right]S = [1 + 1 + 1 - K]S$$

$$= [3 - K]S$$

$$3 - K = d \text{ so } K = 3 - d$$

Fig. 6-4—continued.

Note that this is the *only* value of K that will let the circuit behave properly. Note further that we are once again in the form of expression (A), giving us no interaction. The circuit looks like this:

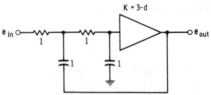

A circuit to provide a high input impedance and a gain of $3 - d$ looks like this:

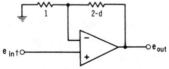

† must return to ground via low-impedance dc path.

The op amp continues to force the $+$ and $-$ inputs to zero. The fraction of the output fed back is $\dfrac{1}{1 + 2 - d} = \dfrac{1}{3 - d}$ so the gain is $3 - d$.

Fig. 6-4—continued.

course, the loading restrictions, hum problems, and cost of the inductor.

The math behind the Sallen-Key sections appears in Fig. 6-4. Using this circuit, there are an incredible variety of tradeoffs between resistance and capacitance ratios, circuit gain, available damping values, and so forth. There are very definite restrictions on what circuit values can go with each other in a Sallen-Key circuit for a certain response. Two of the most useful of the workable Sallen-Key circuits are the *unity-gain Sallen-Key and the equal-component-value Sallen-Key* sections.

Both of these circuits have fixed gains and will operate as expected ONLY at the gain specifically designed into the section. These gains are rather handy, obviously being unity in the unity-gain version and a precisely specified value around 2:1 voltage or +6 dB per second-order section in the equal-component-value case. If we need other gain values, they are easiest to obtain by external adjustment of system levels. Or, if you must, you can attack the math of Fig. 6-4 at your own risk for custom gain values.

The unity-gain Sallen-Key circuits appear in Fig. 6-5. For extreme economy, this circuit can be built with nothing but an emitter follower (Fig. 6-5A), but there are lots of advantages to using an op

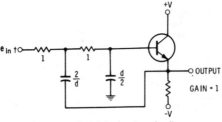

† must return to ground via low-impedance dc path.

(A) Discrete version, normalized to 1 ohm and 1 radian/sec.

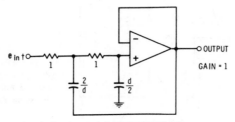

† must return to ground via low-impedance dc path.

(B) Op-amp circuit, normalized to 1 ohm and 1 radian/sec.

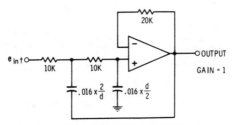

† must return to ground via low-impedance dc path.

(C) Op-amp circuit, normalized to 10K and 1-kHz cutoff frequency.

Fig. 6-5. Simplest form of unity-gain, Sallen-Key, second-order, low-pass, active section.

amp instead. These advantages include a higher ratio of input impedance to output impedance, the absence of a 0.6-volt temperature-dependent offset between input and output, less supply-voltage-to-signal interaction, and, finally, a gain of precisely unity. Unity gain can be important in sections with low damping values. The gain of an emitter follower is always slightly less than unity.

The frequency of this circuit is set by the product of the resistors and capacitors, while the damping is controlled by the *ratio* of the capacitors. The resistors MUST always be identical in value, and the capacitors MUST always be arranged so that the left capacitor is $4/d^2$ times as large as the right one. Fig. 6-6 shows the tuning details.

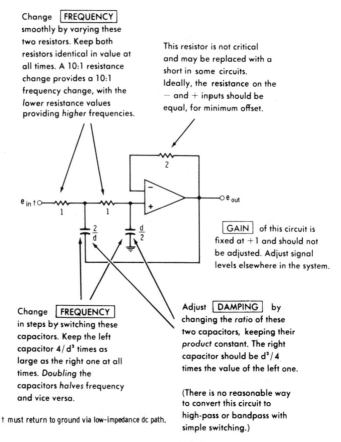

Change FREQUENCY
smoothly by varying these
two resistors. Keep both
resistors identical in value at
all times. A 10:1 resistance
change provides a 10:1
frequency change, with the
lower resistance values
providing *higher* frequencies.

This resistor is not critical
and may be replaced with a
short in some circuits.
Ideally, the resistance on the
− and + inputs should be
equal, for minimum offset.

e_{in} †

GAIN of this circuit is
fixed at +1 and should not
be adjusted. Adjust signal
levels elsewhere in the system.

Change FREQUENCY
in steps by switching these
capacitors. Keep the left
capacitor $4/d^2$ times as
large as the right one at all
times. *Doubling* the
capacitors *halves* frequency
and vice versa.

Adjust DAMPING by
changing the *ratio* of these
two capacitors, keeping their
product constant. The right
capacitor should be $d^2/4$
times the value of the left one.

(There is no reasonable way
to convert this circuit to
high-pass or bandpass with
simple switching.)

† must return to ground via low-impedance dc path.

Fig. 6-6. Adjusting or tuning the unity-gain, Sallen-Key, second-order low-pass section.

Frequency is changed smoothly by changing both resistors simultaneously by means of a dual pot or a ganged pair of trimmers. Frequency is changed in steps by switching capacitors, always keeping their ratio constant at $4/d^2$. A 10:1 change in capacitors provides a decade step in frequency, with larger capacitor values providing lower frequencies.

As with many of the other circuits, the feedback resistor from output to inverting input is not critical and can often be replaced with a short; for minimum offset its optimum value is identical to the dc impedance seen from the noninverting input to ground. We must also provide a dc return path to ground for the noninverting input via the input circuitry.

This circuit is very handy and quite simple, but there are some limitations that may make us look for something better. There is no way that simple switching can be used to convert this circuit to a high-pass circuit of identical performance. The capacitor values are not easy to calculate, and the spread of values gets out of hand for low d values. (A d of 0.1 means a 400:1 capacitor spread.) Finally, damping is hard to trim for minor adjustment without interaction with the frequency to which the section is tuned.

EQUAL-COMPONENT-VALUE SALLEN-KEY CIRCUITS

Few active-filter designers seem aware that there is a "magic" combination of Sallen-Key circuit values that lets everything fall into place for an extremely easy-to-use, easy-to-design, and easy-to-tune filter. This magic combination occurs if we force both resistors to identical values and both capacitors to identical values. This can work with only one value of circuit gain. That magic value is $3 - d$. Very nicely, the gain of the amplifier controls *only* the damping, letting us trim or adjust it at will. The resistors and capacitors are identical in respective values and thus are trivial design considerations. As an extra bonus, it is simple to convert the circuit to an identical high-pass one, just by switching the capacitors and resistors to their opposite positions.

This circuit is called the *equal-component-value Sallen-Key* circuit. It appears in Fig. 6-7. Tuning details are given in Fig. 6-8.

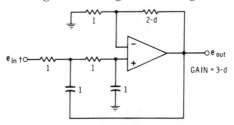

† must return to ground via low-impedance dc path.

(A) Normalized to 1 ohm and 1 radian/sec.

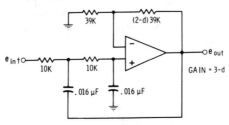

† must return to ground via low-impedance dc path.

(B) Normalized to 10K and 1-kHz cutoff frequency.

Fig. 6-7. Equal-component-value, Sallen-Key, second-order, low-pass filter has independently adjustable damping and frequency.

127

To adjust frequency, adjust both resistors, *keeping their values identical at all times.* The higher resistance values go with the lower frequencies. You can switch frequency in steps by changing capacitors, *always keeping both capacitor values identical.* Capacitance values can be read from Fig. 6-21, or they can be simply calculated as an inverse frequency ratio.

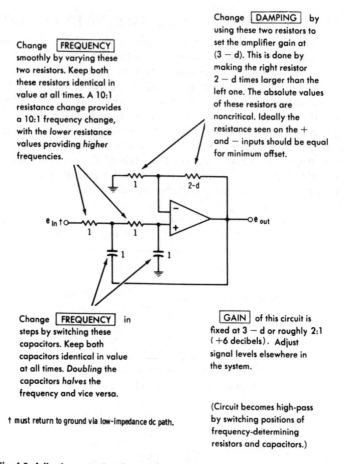

Change DAMPING by using these two resistors to set the amplifier gain at (3 − d). This is done by making the right resistor 2 − d times larger than the left one. The absolute values of these resistors are noncritical. Ideally the resistance seen on the + and − inputs should be equal for minimum offset.

Change FREQUENCY smoothly by varying these two resistors. Keep both these resistors identical in value at all times. A 10:1 resistance change provides a 10:1 frequency change, with the *lower* resistance values providing *higher* frequencies.

Change FREQUENCY in steps by switching these capacitors. Keep both capacitors identical in value at all times. *Doubling* the capacitors *halves* the frequency and vice versa.

† must return to ground via low-impedance dc path.

GAIN of this circuit is fixed at 3 − d or roughly 2:1 (+6 decibels). Adjust signal levels elsewhere in the system.

(Circuit becomes high-pass by switching positions of frequency-determining resistors and capacitors.)

Fig. 6-8. Adjusting or tuning the equal-component-value, Sallen-Key, second-order, low-pass section.

Damping is controlled and adjusted by changing the gain of the amplifier. Note that if we ever attempted to set the gain to a value *greater than 3*, we would have negative damping, or an oscillator. This operating point is well away from normal operation, and the gain is normally stably set by the ratio of two resistors. Nevertheless, if you are going to provide a variable damping control, make sure its highest gain value is something *less* than 3.

The *ratio* of the resistors on the inverting input acts as a voltage divider to set the stage gain and damping. If you are running an op amp very near its upper frequency limit (see Fig. 6-22), damping values slightly lower than normal might be used to compensate for amplifier-gain falloff.

The absolute value of the gain-determining resistors is not particularly critical. Ideally, for minimum offset, their *parallel combination* should equal the total resistance seen to ground at the noninverting input. Thus, with two 10K frequency-determining resistors, the + or noninverting input sees a 20K resistance path to ground. A typical damping resistor calculated as 39K times $(2 - d)$ will be roughly 39K; this in parallel with the 39K divider resistor is about 20K, balancing the impedances on both inputs. Incidentally, with a 39K resistor from inverting input to ground, the critical $d = 0$ feedback resistor would be 78K; feedback resistors always must be *lower* than this value for stable operation.

As usual, we have to provide an op-amp bias path through the input. The circuit is switched to an identically performing high-pass one by using 4pdt (4-pole double-throw) switching to interchange capacitors and resistors.

UNITY-GAIN STATE-VARIABLE CIRCUITS

The equal-component-value Sallen-Key low-pass filter is just about the easiest-to-design and easiest-to-use single-op-amp filter you can possibly obtain, particularly if you have to tune it over a frequency range or have to trim the damping. If you want something still more in the way of performance, you most likely will have to go to a multiple-op-amp universal filter using three, or even four, operational amplifiers.

Why bother? For the majority of fixed or manually tuned low-pass applications, there is no reason to look for anything better, even if extra op amps are cheap. However, there just may be a few times when something better is needed. For instance:

If you need a filter that is easy to tune electronically or control with voltage over a very wide frequency range.

If you need extremely low d values without worrying about stability.

If you need a filter that is very easy to switch to high-pass or bandpass.

If you must have variable gain inside the filter.

If you are involved with things like quadrature art (see Chapter 10), or electronic music where two quadrature outputs (90°-

phase-shifted) are either handy or essential.

If you are working with fancy transfer functions, such as all-pass or bandstop, that need simultaneous outputs.

If you are building a Cauer, or elliptic, filter that needs both a low-pass and a high-pass output at the same time (see Chapter 9).

If any of these advanced requirements are needed, then a universal filter module called the *state-variable* filter is the answer. The state-variable filter uses three or four operational amplifiers. It is shown mathematically in Fig. 6-9.

Despite its fancy name, the circuit is nothing but the analog of a pendulum. Two op amps are connected as inverting integrators in cascade. The output of the second op-amp integrator is unity-gain inverted and sent around to the input of the first integrator. This solves the differential equation for an undamped pendulum or a simple sine-wave oscillator. Additional feedback from the first integrator is also sent back to its own input (adding "rust" or "air resistance" to the "pendulum" or its "hinge") to provide a selected amount of damping.

Finally, an input signal is also routed to the input of the summing stage in front of the first integrator, giving us an electronic input or driving force for the "pendulum." The input-summing stage combines oscillatory feedback, damping, and input signals. With the proper design of this summing block, you can independently adjust the circuit gain, frequency, and damping.

THE MATH BEHIND   **State-variable, low-pass, second-order sections.**

An op-amp integrator circuit looks like this:

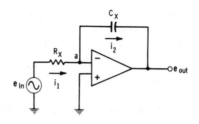

The high gain of the op amp continuously drives the difference between + and − inputs to zero. The voltage on the − input will always be extremely close to ground and may be treated as a *virtual* ground.

Fig. 6-9.

$$i_1 = \frac{e_{in}}{R_x} \text{ since point a is a virtual ground}$$

$$i_2 = \frac{-e_{out}}{1/j\omega C_x} = i_1 = \frac{e_{in}}{R_x}$$

$$\frac{e_{out}}{e_{in}} = -\frac{1}{j\omega R_x C_x} \text{ or letting } S = j\omega$$

$$\frac{e_{out}}{e_{in}} = -\frac{1}{R_x C_x S}$$

The state-variable circuit for analysis looks like this:

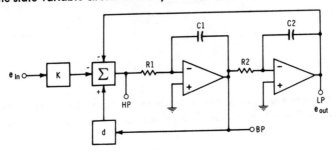

$$e_{hp} = -Ke_{lp} - e_{in} + de_{bp}$$

$$e_{bp} = -\frac{e_{hp}}{SR1C1}$$

$$e_{out} = e_{lp} = -\frac{e_{bp}}{SC2R2} = \frac{e_{hp}}{S^2 R1C1R2C2}$$

$$\therefore \quad e_{hp} = e_{out}(S^2 R1C1R2C2)$$
$$\text{and}$$
$$e_{bp} = -e_{out}(SC2R2)$$

Combining these gives

$$e_{out}(S^2 R1R2C1C2) = -Ke_{in} - e_{out} - d(SC2R2)\dot{e}_{out}$$

$$\boxed{\frac{e_{out}}{e_{in}} = \frac{-K/R1C1R2C2}{S^2 + \dfrac{d}{R1C1}S + \dfrac{1}{R1C1R2C2}}}$$

We see that this is inherently in the form $\dfrac{K}{S^2 + dS + 1}$ for ω or f normal-ized to unity and of form $\dfrac{K\omega^2}{S^2 + d\omega S + \omega^2}$ for any frequency, meaning

Fig. 6-9—continued.

interaction-free adjustment of frequency and damping should be inherent provided we keep R1C1 = R2C2. One problem is to provide a simple summing circuit that allows individual adjustment of gain and damping. A suitable two-op-amp circuit looks like this:

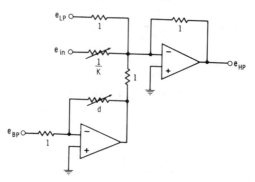

If we attempt to use a single amplifier as a summer, gain-damping interaction will result. A one-amplifier, fixed-gain circuit looks like this:

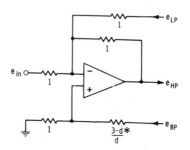

Circuit values are for unity gain. At the + input the gain is 3; the lower two resistors are a voltage divider with a gain of d/3; net gain is +d from the e_bp input. Resistor * may be similarly calculated for other *fixed* gain values.

Fig. 6-9—continued.

There are *three* possible outputs: a low-pass, a bandpass, and a high-pass. All normally have identical gain. At critical damping, the center frequency of the bandpass output equals the cutoff frequencies of the high-pass and low-pass outputs. For a design of d = 0.2, the high-pass and low-pass outputs will behave as filters with a damping value of 0.2; the bandpass section will have a Q of 5, and the peak outputs of all three sections are nominally identical in amplitude and frequency.

We have several possible options for the summing block. A single op amp can sum feedback, an input, and damping in almost any ratio. But, since the damping does not get inverted and the other two signals do, we cannot *independently* adjust the gain—at least not with a single potentiometer—without also changing the damping.

A fixed, unity-gain, state-variable, low-pass filter is shown in Fig. 6-10, and its tuning details are shown in Fig. 6-11. It pays to keep

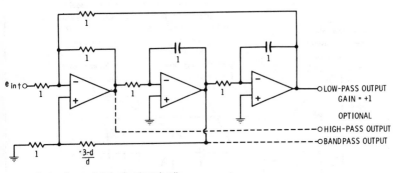

† must return to ground via low-impedance dc path.

(A) Normalized to 1 ohm and 1 radian/sec.

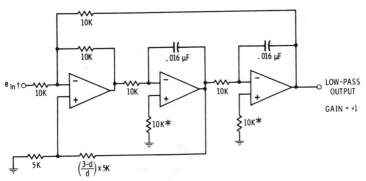

† must return to ground via low-impedance dc path.
* optional offset compensation, may be replaced with short in noncritical circuits.

(B) Normalized to 10K and 1-kHz cutoff frequency.

Fig. 6-10. Three-amplifier, state-variable filter offers low sensitivity, easy voltage-controlled tuning, and easy conversion to bandpass or high-pass. Gain is unity.

the frequency-determining resistors identical in value and the frequency-determining capacitors also identical in value. As with the earlier circuits, varying both resistors together varies frequency inversely, as does step selection of capacitors. The lower frequencies are associated with the larger RC products. The C values are read from Fig. 6-21 or are calculated as simple frequency inverses. Damp-

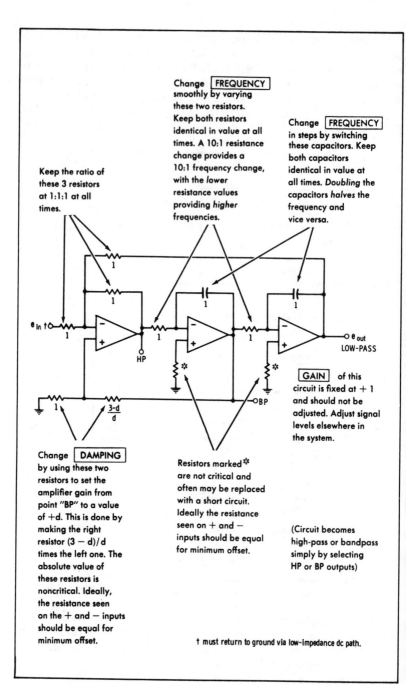

Change **FREQUENCY** smoothly by varying these two resistors. Keep both resistors identical in value at all times. A 10:1 resistance change provides a 10:1 frequency change, with the *lower* resistance values providing *higher* frequencies.

Change **FREQUENCY** in steps by switching these capacitors. Keep both capacitors identical in value at all times. *Doubling* the capacitors *halves* the frequency and vice versa.

Keep the ratio of these 3 resistors at 1:1:1 at all times.

e_{In} ○†

1

1

1

1

1

−

+

HP

1

−

+

✳

1

−

+

✳

○ e_{out}
LOW-PASS

1

$\frac{3-d}{d}$

○ BP

Change **DAMPING** by using these two resistors to set the amplifier gain from point "BP" to a value of +d. This is done by making the right resistor $(3 − d)/d$ times the left one. The absolute value of these resistors is noncritical. Ideally, the resistance seen on the + and − inputs should be equal for minimum offset.

Resistors marked ✳ are not critical and often may be replaced with a short circuit. Ideally the resistance seen on + and − inputs should be equal for minimum offset.

GAIN of this circuit is fixed at + 1 and should not be adjusted. Adjust signal levels elsewhere in the system.

(Circuit becomes high-pass or bandpass simply by selecting HP or BP outputs)

† must return to ground via low-impedance dc path.

Fig. 6-11. Adjusting or tuning the unity-gain, state-variable, second-order low-pass section.

ing is controlled with a single resistor adjustment, calculated as $(3 - d)/d$ times the resistor on the noninverting input of the summing block.

The input, feedback, and summing resistors should be kept at a 1:1:1 ratio for unity gain. As before, we must provide a dc return path for the input through the source circuit. Absolute values of resistors on the + or noninverting inputs once again are not critical and can be used to minimize offset. The ratio of resistors on the noninverting input of the summing op amp should set the overall gain from the bandpass output back through the summer to the high-pass output to a gain value of "d."

Switching to bandpass or high-pass is trivial. Just select the appropriate output with a spdt (single-pole double-throw) or sp3t (single-pole triple-throw) selector switch. You can easily recalculate the input op-amp values for different gains. Remember that the gain on the loop feedback must be −1, and the gain on the "rust" or damping input should be +d. Chapter 2 gave an example of calculation of mixed gains on the inverting and noninverting inputs of a summing amplifier. See Fig. 2-17B.

VARIABLE-GAIN STATE-VARIABLE CIRCUITS

The simplest way to make gain and damping completely independent is to invert the damping signal with a fourth operational amplifier of minus d gain and then sum the input, feedback, and damping signals independently on the inverting input, classic op-amp style. As another benefit, the resistor value for "d" is simply 10K times "d," thus eliminating complex calculations. This variable-gain circuit is shown in Fig. 6-12; tuning and adjustment details are shown in Fig. 6-13. Outside of the gain and damping independence, it is essentially identical to the three-amplifier circuit.

THE "RIPOFF" DEPARTMENT

Stock, Ready-to-Use, Low-pass Active Filters

The equal-component-value Sallen-Key filter is a good basic building block for many filter applications. Figs. 6-14 through 6-20 are a collection of standard, ready-to-use low-pass filters of orders one through six and with seven response options ranging from best-delay through 3-dB dips. (See Chapter 9 for more details on the Cauer, or elliptical, response-shape option.) All of the filters are shown having a 1-kHz cutoff frequency; all of them use identical .016-microfarad capacitors for all points in all stages. To change cutoff frequency, you simply read a new capacitance value from Fig. 6-21 or calculate the ratio of the new capacitor to the old one as the inverse of the frequency ratio. Should the final capacitance value fall between two stock values, just raise or lower the impedance of the circuit to hit the right value.

All resistance values needed have been rounded off to stock one-percent values. The guidelines of actual accuracies needed appear in Fig. 4-19, while each of the charts indicates a recommended working tolerance for each circuit. Five-percent values are more than ade-

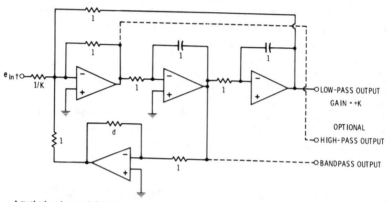

(A) Normalized to 1 ohm and 1 radian/sec.

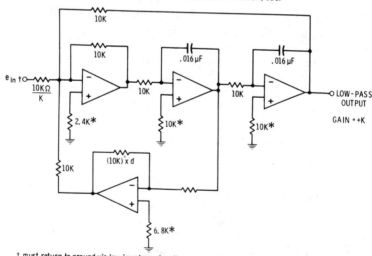

(B) Normalized to 10K and 1-kHz cutoff frequency.

Fig. 6-12. Variable-gain, state-variable filter. Gain, frequency, and damping are independently adjustable.

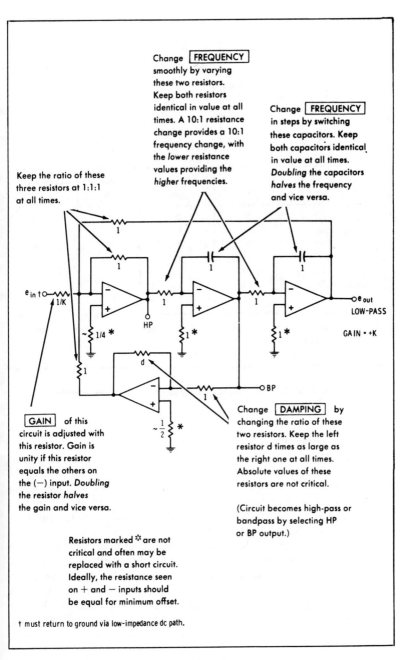

Change **FREQUENCY** smoothly by varying these two resistors. Keep both resistors identical in value at all times. A 10:1 resistance change provides a 10:1 frequency change, with the *lower* resistance values providing the *higher* frequencies.

Change **FREQUENCY** in steps by switching these capacitors. Keep both capacitors identical in value at all times. *Doubling* the capacitors *halves* the frequency and vice versa.

Keep the ratio of these three resistors at 1:1:1 at all times.

e_{in} †

HP

e out
LOW-PASS

GAIN = +K

BP

GAIN of this circuit is adjusted with this resistor. Gain is unity if this resistor equals the others on the (—) input. *Doubling* the resistor *halves* the gain and vice versa.

Change **DAMPING** by changing the ratio of these two resistors. Keep the left resistor d times as large as the right one at all times. Absolute values of these resistors are not critical.

(Circuit becomes high-pass or bandpass by selecting HP or BP output.)

Resistors marked ✻ are not critical and often may be replaced with a short circuit. Ideally, the resistance seen on + and — inputs should be equal for minimum offset.

† must return to ground via low-impedance dc path.

Fig. 6-13. Adjusting or tuning the variable-gain, state-variable, second-order low-pass section.

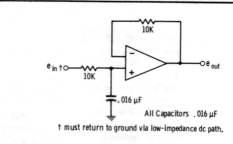

Response	RF1	Gain e_{out}/e_{in}	Gain Decibels	Component Tolerance
Best Delay	10K	1	0	10%
Compromise	10K	1	0	10%
Flattest Amp	10K	1	0	10%
Slight Dips	10K	1	0	10%
1-Decibel Dip	10K	1	0	10%
2-Decibel Dip	10K	1	0	10%
3-Decibel Dip	10K	1	0	10%

To change frequency, scale all capacitors suitably. *Tripling* the capacity cuts frequency by one-third, and vice versa.

Fig. 6-14. First-order, low-pass circuits, —6 dB/octave rolloff, 1-kHz cutoff frequency.

quate for the majority of the circuits. Nevertheless, use the best accuracy you possibly can.

The filters are arranged with the higher-damped sections toward the input. The overall circuit gain is also given, both as an output/input ratio and in decibels. To use one of these, first decide which one you want, using Chapter 4 as a guide. Then draw its schematic, substitute the resistance values, and scale the capacitors to your cutoff frequency. That's all there is to it.

First-, second-, and third-order circuits appear in Figs. 6-14 through 6-16. The second op amp can usually be eliminated from a noncritical third-order filter by using the trick of Fig. 6-17. Here, the input RC section is unloaded by reducing its impedance to one-tenth of normal, so the following second-order section does not load it excessively. This drops the input impedance to around 1000 ohms, compared to the 10K or so nominal impedance of the second-order section.

Fourth-, fifth, and sixth-order filters with rolloff responses of 24, 30, and 36 dB per octave appear in Figs. 6-18 through 6-20. Use of a

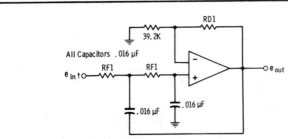

All Capacitors .016 µF

† must return to ground via low-impedance dc path.

Response	RF1	RD1	Gain e_{out}/e_{in}	Gain Decibels	Component Tolerance
Best Delay	7.87K	10.5K	1.3	2.3	10%
Compromise	8.87K	16.9K	1.4	3.0	10%
Flattest Amp	10.0K	22.6K	1.6	4.1	10%
Slight Dips	10.7K	30.9K	1.8	5.2	10%
1-Decibel Dip	11.5K	37.4K	2.0	6.0	10%
2-Decibel Dip	11.8K	43.2K	2.1	6.4	10%
3-Decibel Dip	11.8K	48.7K	2.2	6.8	5%

To change frequency, scale all capacitors suitably. *Tripling* the capacity cuts frequency by *one-third*, and vice versa.

Fig. 6-15. Second-order, low-pass circuits, —12 dB/octave rolloff, 1-kHz cutoff frequency.

two-amplifier fifth-order filter similar to that of Fig. 6-17 is possible, but some trimming will most likely be needed.

Capacitor values for different frequencies are calculated using Fig. 6-21.

PITFALLS AND RESTRICTIONS

Fig. 6-22 shows the recommended upper-frequency limits for the 741 and LM318 operational amplifiers used in the four basic filter types of this chapter. Whenever you are operating near the recommended limit, damping values may have to be decreased slightly to get the desired responses.

Remember that slew-rate limitations may place a much more severe restriction on operating frequency when you are dealing with high-amplitude, high-frequency signals. This was outlined in Chapter 2.

A list of some of the other pitfalls and design problems you may encounter appears in Fig. 6-23. The most common of these are fail-

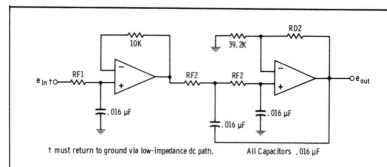

Response	RF1	RF2	RD2	Gain e_{out}/e_{in}	Gain Decibels	Component Tolerance
Best Delay	7.50K	6.81K	21.5K	1.6	4.1	10%
Compromise	8.68K	8.25K	31.6K	1.8	5.1	10%
Flattest Amp	10.0K	10.0K	39.2K	2.0	6.0	10%
Slight Dips	15.0K	10.5K	51.1K	2.3	7.3	10%
1-Decibel Dip	22.1K	11.0K	59.0K	2.5	8.0	5%
2-Decibel Dip	30.9K	11.0K	63.4K	2.6	8.3'	5%
3-Decibel Dip	33.2K	11.0K	66.5K	2.7	8.6	2%

To change frequency, scale all capacitors suitably. *Tripling* the capacity cuts frequency by *one-third,* and vice versa.

Fig. 6-16. Third-order, low-pass circuits, —18 dB/octave rolloff, 1-kHz cutoff frequency.

ing to provide an input bias dc return path and ignoring the very specific capacitor and resistor ratios that are called for in the various circuits.

SOME DESIGN RULES

We can summarize the design rules very simply:

If you can use the equal-component-value Sallen-Key circuit:

1. Referring to your original filter problem and using Chapter 4, select a shape and order that will do the job.
2. Construct this circuit from Figs. 6-14 through 6-20 and substitute the proper resistance values.
3. Scale the circuit to your cutoff frequency by using Fig. 6-21 or by calculating capacitor ratios inversely as frequency.
4. Tune and adjust the circuit by using the guidelines in this chapter and Chapter 9. For very low frequencies, consider a 10X increase in impedance level to get by with smaller capacitors.

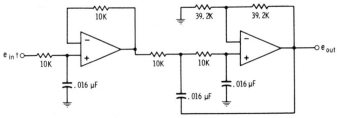

† must return to ground via low-impedance dc path.

(A) Typical third-order, low-pass with two op amps.

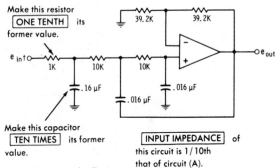

Make this resistor ONE TENTH its former value.

Make this capacitor TEN TIMES its former value.

INPUT IMPEDANCE of this circuit is 1/10th that of circuit (A).

† must return to ground via low-impedance dc path.

(B) One-op-amp approximation to (A).

Fig. 6-17. Approximating a third-order, low-pass with a single op amp.

To build any low-pass active filter:

1. Referring to your original filter problem and using Chapter 4, select a shape and order that will do the job, including a list of the frequencies and damping values for each section to be cascaded, along with an accuracy specification.

2. Select a suitable second-order section from this chapter for each section you need, normalized to 1 kHz. Shift the frequencies of the cascaded sections as called for to realize the particular shape. Remember that *increasing* the resistance or capacitance *decreases* the frequency.

3. Set the damping value of each cascaded section as called for.

4. Scale the circuit to your cutoff frequency by using Fig. 6-21 or by calculating capacitors inversely as frequency.

5. Arrange the circuits starting with the highest-damped one first, adding a first-order active section if needed.

6. Tune and adjust the circuit, using the guidelines of this chapter and Chapter 9. For very low frequencies, consider a 10X increase in impedance level to get by with smaller capacitors.

Several low-pass filter design examples appear in Fig. 6-24.

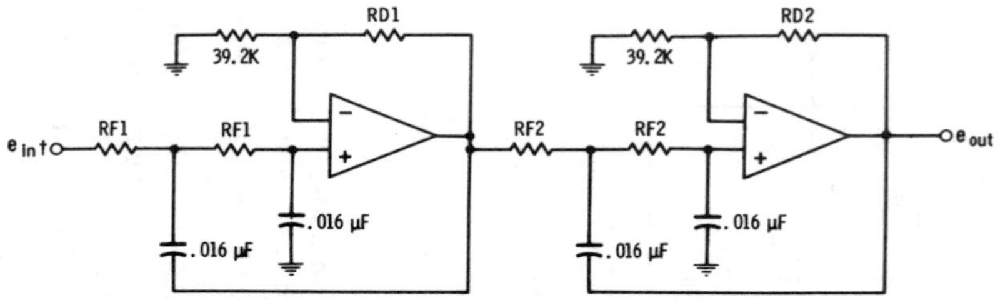

All Capacitors .016 μF

† must return to ground via low-impedance dc path.

Response	RF1	RD1	RF2	RD2	Gain e_{out}/e_{in}	Gain Decibels	Component Tolerance
Best Delay	6.98K	3.24K	6.19K	29.4K	+1.90	5.6	10%
Compromise	8.25K	4.64K	7.87K	41.2K	+2.30	7.2	10%
Flattest Amp	10.0K	5.90K	10.0K	48.7K	+2.60	8.3†	5%
Slight Dips	14.0K	18.2K	10.2K	60.4K	+3.72	11.4	5%
1-Decibel Dip	19.1K	28.7K	10.5K	66.5K	+4.70	13.4	5%
2-Decibel Dip	21.5K	35.7K	10.5K	69.8K	+5.31	14.5	2%
3-Decibel Dip	22.6K	42.2K	10.5K	71.5K	+5.84	15.3	1%

To change frequency, scale all capacitors suitably. Tripling the capacity cuts frequency by one-third, and vice versa.

Fig. 6-18. Fourth-order low-pass circuits, −24 dB/octave rolloff, 1-kHz cutoff frequency.

142

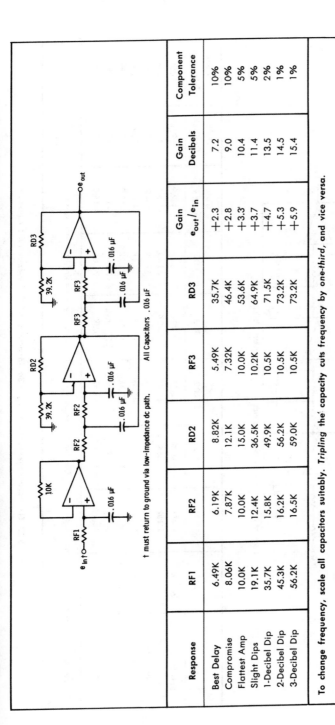

Response	RF1	RF2	RD2	RF3	RD3	Gain e_{out}/e_{in}	Gain Decibels	Component Tolerance
Best Delay	6.49K	6.19K	8.82K	5.49K	35.7K	+2.3	7.2	10%
Compromise	8.06K	7.87K	12.1K	7.32K	46.4K	+2.8	9.0	10%
Flattest Amp	10.0K	10.0K	15.0K	10.0K	53.6K	+3.3	10.4	5%
Slight Dips	19.1K	12.4K	36.5K	10.2K	64.9K	+3.7	11.4	5%
1-Decibel Dip	35.7K	15.8K	49.9K	10.5K	71.5K	+4.7	13.5	2%
2-Decibel Dip	45.3K	16.2K	56.2K	10.5K	73.2K	+5.3	14.5	1%
3-Decibel Dip	56.2K	16.5K	59.0K	10.5K	73.2K	+5.9	15.4	1%

To change frequency, scale all capacitors suitably. *Tripling the capacity cuts frequency by one-third, and vice versa.*

Fig. 6-19. Fifth-order low-pass circuits, −30 dB/octave rolloff, 1-kHz cutoff frequency.

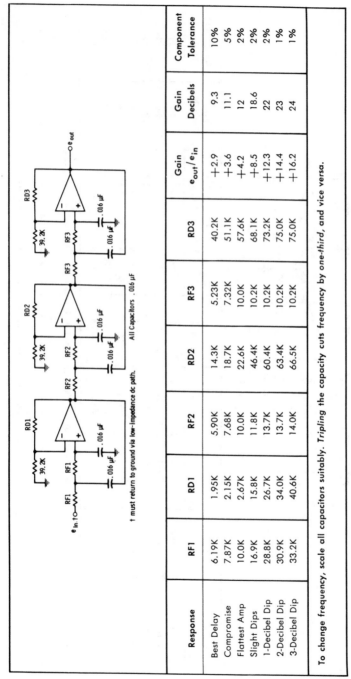

† must return to ground via low-impedance dc path.

All Capacitors .016 μF

Response	RF1	RD1	RF2	RD2	RF3	RD3	Gain e_{out}/e_{in}	Gain Decibels	Component Tolerance
Best Delay	6.19K	1.95K	5.90K	14.3K	5.23K	40.2K	+2.9	9.3	10%
Compromise	7.87K	2.15K	7.68K	18.7K	7.32K	51.1K	+3.6	11.1	5%
Flattest Amp	10.0K	2.67K	10.0K	22.6K	10.0K	57.6K	+4.2	12	2%
Slight Dips	16.9K	15.8K	11.8K	46.4K	10.2K	68.1K	+8.5	18.6	2%
1-Decibel Dip	28.8K	26.7K	13.7K	60.4K	10.2K	73.2K	+12.3	22	2%
2-Decibel Dip	30.9K	34.0K	13.7K	63.4K	10.2K	75.0K	+14.4	23	1%
3-Decibel Dip	33.2K	40.6K	14.0K	66.5K	10.2K	75.0K	+16.2	24	1%

To change frequency, scale all capacitors suitably. *Tripling the capacity cuts frequency by one-third,* and vice versa.

Fig. 6-20. Sixth-order low-pass circuits, —36 dB/octave rolloff, 1-kHz cutoff frequency.

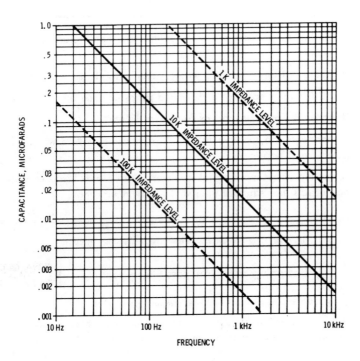

Fig. 6-21. Capacitance values for frequency scaling.

	741	LM318
Unity-Gain Sallen-Key	25 kHz	500 kHz
Equal-Component-Value Sallen-Key	10 kHz	200 kHz
Unity-Gain State-Variable	25 kHz	500 kHz
Gain-of-Ten State-Variable	2.5 kHz	50 kHz

Fig. 6-22. Recommended upper cutoff frequency limits for the op amps of Chapter 3.

1. Forgetting to provide a low-impedance dc input return path to ground.

2. Damping resistors missing or wrong value or too loose a tolerance.

3. R and C values not kept as specified ratios of each other.

4. Using an op amp beyond its frequency limits.

5. Components too loose in tolerance or tracking poorly when adjusted (see Chapter 9).

6. Forgetting about input signal levels or nonzero bias levels or voltage offsets that saturate filter or limit dynamic range.

Fig. 6-23. Common pitfalls in low-pass, active-filter circuits.

Designing active low-pass filters—some **EXAMPLES**

A. Design a 250-hertz, third-order, 1-dB dip, low-pass, active filter.

Use the circuit of Fig. 6-16, placing a 22.1K resistor on the first-order stage, an 11.0K resistor for the second-order frequency resistors, and a 59.0K resistor to set the second-stage damping. The capacitor values for a 250-Hz frequency will be 1000/250 of their 1-kHz value or .062 µF, either as calculated or as read from Fig. 6-21. The circuit looks like this:

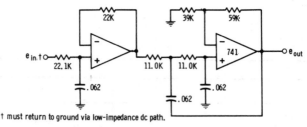

† must return to ground via low-impedance dc path.

Or, we can eliminate the first op amp by reducing the input resistor to 2.21K and increasing the capacitor to 0.62 µF:

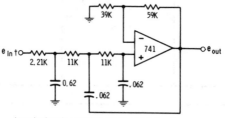

Fig. 6-24.

† must return to ground via low-impedance dc path.

Ten-percent components are acceptable for either circuit.

B. A filter for a biomedical experiment has to have a cutoff frequency of 10 Hz, a tolerable transient and overshoot response, and must reject frequencies above 15 Hz by a minimum of 30 dB. Design the filter.

Fifteen Hz is 1.5 times the 10-hertz cutoff frequency. From Fig. 4-8A, we see that a 1-, 2-, or 3-dB dips fifth-order filter will work, while from Fig. 4-9A, a sixth-order slight-dips filter will also do the job. Looking at the damping values in Figs. 4-8 and 4-9, we conclude that we will probably get the best transient performance with the sixth-order slight-dips filter. We then go to Fig. 6-20 and select the component values. Since 10 Hz is 1/100 times 1 kHz, the capacitance values have to be 100 times the normal values, or 1.6 microfarads. This is a bit large, so, let's scale impedance by a factor of 10, multiplying all resistors by 10 and dividing all capacitors by ten. The final circuit looks like this:

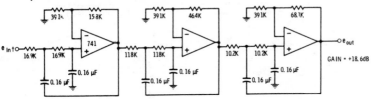

† must return to ground via low-impedance dc path.

Two percent tolerance is recommended for this circuit. If we are interested only in rejecting 15 Hz, the elliptical techniques of Chapter 9 may use simpler circuits.

C. Part of a precision telephone-network active equalizer needs a second-order low-pass section with a cutoff frequency of 45 kHz and a damping value of 0.082. Design the section.

A precision application with a low damping suggests a state-variable filter, and the high frequency demands a premium op amp such as the LM318. We use the circuit of Fig. 6-10. Capacitors are scaled to 1/45 their normal value to raise the frequency to 45 kHz, leaving us with 355 pF. The damping resistor is calculated as (3 — d)/d times 5K or 178K. The section looks like this:

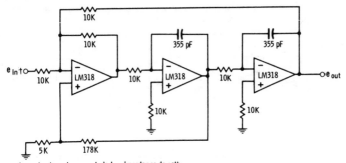

† must return to ground via low-impedance dc path.

Component tolerance would depend on the application; for such a low damping value, 1% components may be needed.

Fig. 6-24—continued.

D. Design a general-purpose, fourth-order, flattest-amplitude, variable lab filter, adjustable from 10 Hz to 10 kHz.

We will do this problem two ways since we will need the results in a later example. The maximum-flatness filter is often the best choice when it has to be tuned, as all the frequency-determining resistors are equal for all stages, making wide-range tuning easy and still allowing identical stock capacitors on each stage.

Using equal-component-value Sallen-Key:

From Fig. 6-18, we let the frequency-determining resistors vary from 10K to 110K, using a quad 100K potentiometer (See Chapter 9 for more details on this technique). Damping resistors are 5.90K and 48.7K. Capacitors are switched in value. .016μF and 10K will be the upper end of the 10-kHz range, 0.16 μF for the 100-Hz range, and 1.6 μF for the 10-Hz range. The circuit looks like this:

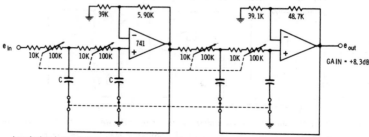

† must return to ground via low-impedance dc path.

RANGE	C
1 - 10 Hz	1.600 μF
10 - 100 Hz	0.160 μF
100 - 1 kHz	0.016 μF
1 - 10 kHz	1600 pF

Using state-variable sections:

We use Fig. 4-7B. Damping values we need are 1.848 for the first section and 0.765 for the second. The equivalent $(3 - d)/d$ times 5K resistance values are 14.7K and 3.09K. The rest of the design is essentially the same, and the final circuit looks like this:

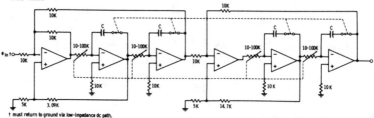

† must return to ground via low-impedance dc path.

Five-percent component values are recommended for either circuit. We will see in Chapter 8 that either of these circuits readily converts to a high-pass one, although the equal-component-value circuit needs 8-pole, double-throw switching, while the state-variable filter only needs a simple double-pole, double-throw switch.

Fig. 6-24—continued.

CHAPTER 7

Bandpass Filter Circuits

Chapter 5 showed us how to take the need for a bandpass response and convert this need into the specifications for a cascaded group of one, two, or three resonant poles of a certain center frequency, staggering, and Q. In this chapter, we will discuss how to build the actual circuits we need, starting with these specifications. After picking up the circuits and learning how to tune them, we will look at a few examples. We will end with an often neglected topic, the transient decay characteristics of high-Q filters, that is becoming very important for electronic-music percussion applications.

Bandpass circuits are normally associated with much lower damping and higher Q values than the usual low- or high-pass responses. We saw that the low-pass filters took tighter and tighter tolerances and more-stringent gain restrictions as damping was lowered, which made the circuits progressively harder to build and tune as the damping went down and the Q went up. This is also true of bandpass circuits.

In fact, it has pretty much been proven that a high-performance, high-Q bandpass active pole can NOT be built with a single operational amplifier. With a single op-amp circuit, you will ALWAYS get into component-spread problems, sensitivity problems, or severe gain restrictions as you try to raise the circuit Q beyond a certain point.

If we are only interested in low-Q applications in the $Q = 2$ to $Q = 5$ range, we have a choice of several circuits, some of them similar to the single op-amp low-pass versions of the last chapter. For Q values of 50 to 500 or more, we MUST use three and four op-amp circuits to get a stable, useful response. For intermediate Q values, the multiple-op-amp circuits are strongly recommended, particularly if wide tuning is needed. Fig. 7-1 sums up these recommendations.

Fortunately, Q values in the 2 to 5 range are ideal for many audio problems, including equalizers, tone modifiers, formant filters for electronic music, psychedelic lighting systems, and so on (see Chapter 10). On the other hand, asking too much Q or too high a frequency from these ultrasimple circuits is asking for problems. Single op-amp bandpass filters should only be used for low-Q applications where a bunch of them are used together in multiple channels. The 3- and 4-op-amp circuits should be used for all other needs.

Our main stock in trade will be the *multiple-feedback bandpass filter,* a single IC circuit and two *state-variable bandpass filters,* a 3-amplifier fixed-gain variation, and a 4-amplifier variable-gain version. We will also take a brief look at some *Sallen-Key* single-amplifier circuits and a *biquad* 3-amplifier filter. These latter versions are occasionally used for special purposes.

MULTIPLE-FEEDBACK BANDPASS CIRCUIT

Fig. 7-2 shows two versions of this single-IC bandpass section. One section builds a single resonant pole equivalent to a second-order factor of the bandpass response we are after.

There are two feedback loops in the circuits. The 2Q resistor from op-amp output to input sets the gain and the current through the lower, frequency-determining capacitor. The upper capacitor provides feedback from the output to the middle of the circuit. The math involved appears in Fig. 7-3.

---

1. Use the methods of this chapter only if the percentage bandwidth is less than 80 to 100%. For wider bandwidths, use overlapping high-pass and low-pass filters instead.

2. You can NOT build a stable, high-Q, easy-to-tune, single-IC bandpass filter.

   —Single-IC circuits may be used for Qs in the 2 to 5 range.

   —Multiple-IC circuits (state-variable or biquad) *must* be used for Qs of 25 to 500.

   —For intermediate Q values and where wide tuning is needed, multiple-IC circuits are strongly urged.

Fig. 7-1. Guidelines for bandpass filter circuits.

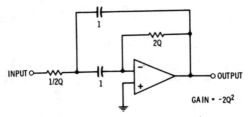

(A) Normalized to 1 ohm and 1 radian/sec.

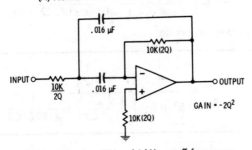

(B) Normalized to 10K and 1-kHz cutoff frequency.

Fig. 7-2. Single-amplifier, multiple-feedback, bandpass circuit.

| THE MATH BEHIND | The single-amplifier, multiple-feedback, bandpass filter. |

Fig. 7-2 may be redrawn for analysis:

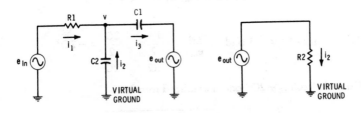

Solving for currents and summing:

$$i_2 = \frac{e_{out}}{R2}v = -i_2 Z_{C2} = \frac{-i_2}{j\omega C2} = \frac{-e_{out}}{j\omega R2 C2}$$

$$i_1 = \frac{e_{in} - v}{R1} = \frac{e_{in}}{R1} + \frac{e_{out}}{j\omega R1 R2 C2}$$

Fig. 7-3.

$$i_3 = \frac{v - e_{out}}{Z_{C1}} = (v - e_{out})j\omega C1 = -e_{out}j\omega C1\left[1 + \frac{1}{j\omega R2C2}\right]$$

$$i_1 + i_2 = i_3$$

$$\frac{e_{in}}{R1} + \frac{e_{out}}{j\omega R1R2C2} + \frac{e_{out}}{R2} = -e_{out}j\omega C1\left[1 + \frac{1}{j\omega R2C2}\right]$$

This rearranges to

$$\frac{e_{out}}{e_{in}} = \frac{-j\omega R2C2}{1 + j\omega R1C2 + j\omega C1(j\omega R1R2C2 + R1)}$$

Letting $S = j\omega$ and reworking

$$\frac{e_{out}}{e_{in}} = \frac{-S\dfrac{1}{R1C1}}{S^2 + S\dfrac{1}{R2}\left(\dfrac{1}{C1} + \dfrac{1}{C2}\right) + \dfrac{1}{R1R2C1C2}}$$

ω will equal $\sqrt{\dfrac{1}{R1R2C1C2}}$. Let $C1 = C2 = 1$. $R1 = 1/R2$.

For equal C values, at $\omega = 1$

$$\frac{e_{out}}{e_{in}} = \frac{-S\left(\dfrac{1}{R1}\right)}{S^2 + \dfrac{2}{R2}S + 1}.$$

Let $R2 = 2Q$. $R1 = 1/2Q$

At resonance $S^2 = -1$ and $= \dfrac{e_{out}}{e_{in}} = -\dfrac{\dfrac{1}{R1}}{\dfrac{2}{R2}} = -\dfrac{2Q}{\dfrac{2}{2Q}} = -2Q^2$.

Gain of circuit is $-2Q^2$ and normalized form is

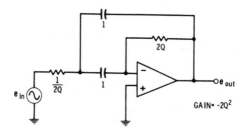

GAIN = $-2Q^2$

Fig. 7-3—continued.

152

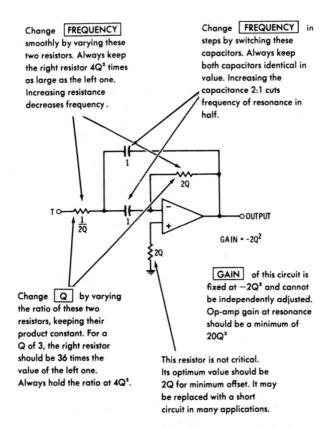

Change [FREQUENCY] smoothly by varying these two resistors. Always keep the right resistor $4Q^2$ times as large as the left one. Increasing resistance decreases frequency.

Change [FREQUENCY] in steps by switching these capacitors. Always keep both capacitors identical in value. Increasing the capacitance 2:1 cuts frequency of resonance in half.

GAIN = $-2Q^2$

Change [Q] by varying the ratio of these two resistors, keeping their product constant. For a Q of 3, the right resistor should be 36 times the value of the left one. Always hold the ratio at $4Q^2$.

[GAIN] of this circuit is fixed at $-2Q^2$ and cannot be independently adjusted. Op-amp gain at resonance should be a minimum of $20Q^2$

This resistor is not critical. Its optimum value should be 2Q for minimum offset. It may be replaced with a short circuit in many applications.

Fig. 7-4. Tuning the single-amplifier, multiple-feedback circuit.

This is a simple and well-behaved circuit at low to moderate Q. The gain is $-2Q^2$, and the op amp has to have an open-loop gain of at least $20Q^2$ at the resonance frequency. So as Q goes up, the useful maximum resonance frequency drops dramatically. (See Fig. 7-15.) The spread of resistor values also increases to the tune of $4Q^2$, thus lowering the input resistance and raising the op-amp feedback resistance.

The circuit is tuned by following Fig. 7-4. We can switch capacitors, providing a 10:1 change in frequency with a 10:1 capacitor change, but we have to keep both capacitors identical in value at all times. We can vary the resistance as well, but we must keep the right resistor $4Q^2$ times the value of the left one at all times. The limits to resistance tuning are set by the maximum value the op-amp feedback resistor can be and the minimum value we are willing to drive. The offset of the op amp will also change as we change the feedback resistor.

We can change the Q and the bandwidth by changing the *ratio* of the two tuning resistors, keeping their product constant. As we change frequency, the Q and the percentage bandwidth stay constant. Thus, for higher frequencies, the bandwidth automatically increases to always give us the same percentage we selected.

Fig. 7-5 shows how we can add a resistor to the input. This will lower the circuit gain and raise the input impedance at the same time. For instance, with a 10:1 input attenuation, the input impedance will be raised by almost 11 times, and the gain will be cut by a factor of 10, from $-2Q^2$ to $-Q^2/5$. Note that this resistor does nothing to the frequency limits on the op amp; we still need an open-loop gain of $20Q^2$ at the resonance frequency.

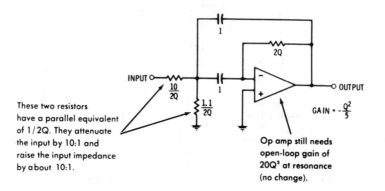

These two resistors have a parallel equivalent of $1/2Q$. They attenuate the input by 10:1 and raise the input impedance by about 10:1.

Op amp still needs open-loop gain of $20Q^2$ at resonance (no change).

$$GAIN = -\frac{Q^2}{5}$$

Fig. 7-5. Modified version of multiple-feedback circuit reduces gain, raises input impedance.

This is a good, general-purpose, low-Q circuit. The upper Q limit depends on the op amp, the frequency, and the type of component spread you can drive. While Qs of 50 and above are possible, the circuit works best for Qs less than 10.

SALLEN-KEY BANDPASS CIRCUIT

We can build Sallen-Key bandpass circuits as well, but they have very serious tuning and frequency restrictions. Fig. 7-6A shows the basic one-amplifier circuit. It has the advantage of using very small capacitors, making the circuit potentially useful for very low frequency work. Circuit gain is $-3Q$, but the op amp must have a gain of at least $90Q^2$ at the center frequency, and there is a strong interaction between frequency and Q. The gain restrictions are lifted somewhat by using two amplifiers, as shown in Fig. 7-6B, but tuning problems still remain.

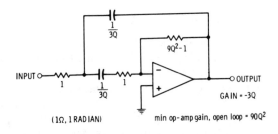

(1Ω, 1 RADIAN)

GAIN = -3Q

min op-amp gain, open loop = $90Q^2$

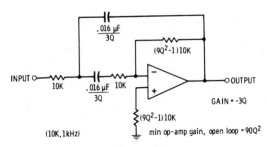

(10K, 1kHz)

GAIN = -3Q

min op-amp gain, open loop = $90Q^2$

(A) Single amplifier.

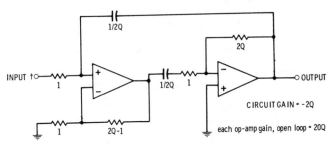

CIRCUIT GAIN = -2Q

each op-amp gain, open loop = 20Q

† must return to ground via low-impedance dc path.

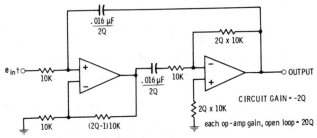

CIRCUIT GAIN = -2Q

each op-amp gain, open loop = 20Q

† must return to ground via low-impedance dc path.

(B) Double amplifier.

Fig. 7-6. Sallen-Key bandpass circuits (see text).

STATE-VARIABLE ACTIVE FILTERS

We saw in the last chapter how the state-variable filter using three or four IC op amps is a universal filter with three outputs: a low-pass, a bandpass, and a high-pass. We can even sum these outputs for fancier second-order responses. The state-variable filter consists of two cascaded integrators and a summing block that combines input and feedback signals in the proper ratio to get a desired response. The circuit is essentially an analog computer model of a transfer function equivalent to the one that we are after.

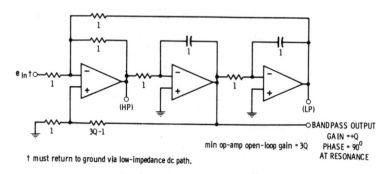

(A) Normalized to 1 ohm and 1 radian/sec.

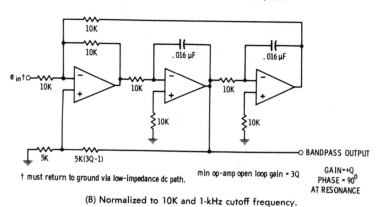

(B) Normalized to 10K and 1-kHz cutoff frequency.

Fig. 7-7. Gain of a Q state-variable bandpass circuit.

Fig. 7-7 shows the 3-amplifier circuit. This time, we have connected it for bandpass use and expressed the feedback resistor in terms of Q rather than of damping. The circuit gain is +Q, and the op amp need only have a gain of 3Q or so, open-loop, at the resonance frequency. This is a much less severe restriction than the earlier filters, so state-variable techniques are ideal for high-Q and high-frequency uses, for a particular choice of IC. The math behind this circuit appears in Fig. 7-8.

Fig. 7-9 shows how we can tune the filter. As before, we switch capacitors for step changes in frequency, keeping their values identical. Two frequency-determining resistors can also be simultaneously varied to change frequency independent of Q or gain. Circuit Q and bandwidth are adjusted with a single resistor. As the frequency changes, the Q and percentage bandwidth stay constant. The abso-

THE MATH BEHIND   State-variable bandpass second-order sections.

An op-amp integrator looks like this:

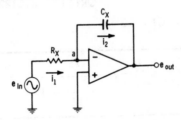

The high gain of the op amp continuously drives point a to ground, forming a *virtual ground*.

$$i_1 = \frac{e_{in}}{R_x} \text{ since point a is a virtual ground.}$$

$$i_2 = \frac{-e_{out}}{1/j\omega C_x} = i_1 = \frac{e_{in}}{R_x} \qquad \frac{e_{out}}{e_{in}} = -\frac{1}{j\omega R_x C_x}$$

or letting $S = j\omega$

$$\frac{e_{out}}{e_{in}} = -\frac{1}{R_x C_x S}$$

the state-variable circuit for analysis looks like this:

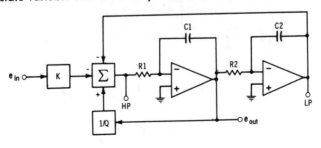

Fig. 7-8.

$$e_{lp} = e_{out}\left[-\frac{1}{R2C2S}\right]$$

$$e_{hp} = -e_{lp} - Ke_{in} + \frac{1}{Q}e_{out} = \left[\frac{1}{R2C2S} + \frac{1}{Q}\right]e_{out} - Ke_{in}$$

$$e_{out} = e_{hp}\left[-\frac{1}{R1C1S}\right] = \left[R1C1S + \left(\frac{1}{R2C2S} + \frac{1}{Q}\right)e_{out} - Ke_{in}\right]$$

Simplifying yields

$$\frac{e_{out}}{e_{in}} = K\frac{\dfrac{S^2}{R1C1}}{S^2 + \dfrac{1}{QR1C1}S + \dfrac{1}{R1C1R2C2}}$$

or letting $R1 = R2 = C1 = C2$

$$\frac{e_{out}}{e_{in}} = K\frac{S^2}{S^2 + \dfrac{1}{Q}S + 1}$$

At $S = 1$ $S^2 = -1$

and $\dfrac{e_{out}}{e_{in}} = KQ$

The summing block can be realized several ways:
(A) Q gain

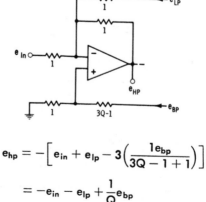

$$e_{hp} = -\left[e_{in} + e_{lp} - 3\left(\frac{1e_{bp}}{3Q - 1 + 1}\right)\right]$$

$$= -e_{in} - e_{lp} + \frac{1}{Q}e_{bp}$$

Fig. 7-8—continued.

158

(B) Variable gain

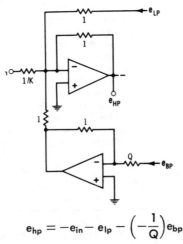

$$e_{hp} = -e_{in} - e_{lp} - \left(-\frac{1}{Q}\right)e_{bp}$$

Fig. 7-8—continued.

lute bandwidth goes up or down proportionately with the center frequency.

One more op amp gives us variable gain, independent of frequency and bandwidth. Details are shown in Fig. 7-10, and the circuit is tuned as shown in Fig. 7-11.

There is one refinement we might like to add to bandpass versions of the state-variable filter. The Q resistors can get rather large with respect to the other components. Circuit strays, particularly the op-amp input capacitance, can cause shifts in response. Fig. 7-12 shows how to use a voltage divider on the bandpass output to lower the value of the Q-determining network. All that we have really done is replace one high-value resistor with three lower-value ones having the same equivalent current-proportioning ratio.

Note that at resonance the output phase lags the input by 90 degrees, or is in *quadrature*. Note also that the gain is +Q for the 3-amplifier circuit. The high gain limits the maximum size of input signal that can be allowed. For instance, with a 5-volt peak-to-peak output swing and a Q of 100, the input signal has to be limited to 50 millivolts or less to keep from saturating the amplifier. Care should also be taken at high gain values to keep input and output signal paths well separated.

THE BIQUAD

Fig. 7-13 shows a circuit that resembles a state-variable circuit, but only in appearance. It is called a *biquad*. The circuit consists of

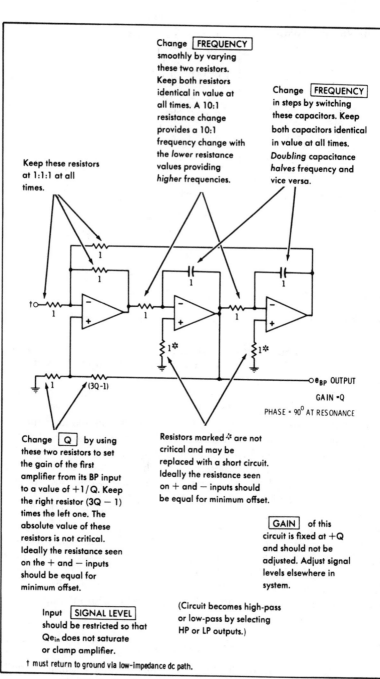

Change [FREQUENCY] smoothly by varying these two resistors. Keep both resistors identical in value at all times. A 10:1 resistance change provides a 10:1 frequency change with the *lower* resistance values providing *higher* frequencies.

Change [FREQUENCY] in steps by switching these capacitors. Keep both capacitors identical in value at all times. *Doubling* capacitance *halves* frequency and vice versa.

Keep these resistors at 1:1:1 at all times.

oe_{BP} OUTPUT

GAIN $=Q$

PHASE $= 90^0$ AT RESONANCE

Change [Q] by using these two resistors to set the gain of the first amplifier from its BP input to a value of $+1/Q$. Keep the right resistor $(3Q - 1)$ times the left one. The absolute value of these resistors is not critical. Ideally the resistance seen on the + and − inputs should be equal for minimum offset.

Resistors marked ✷ are not critical and may be replaced with a short circuit. Ideally the resistance seen on + and − inputs should be equal for minimum offset.

[GAIN] of this circuit is fixed at $+Q$ and should not be adjusted. Adjust signal levels elsewhere in system.

Input [SIGNAL LEVEL] should be restricted so that Qe_{in} does not saturate or clamp amplifier.

(Circuit becomes high-pass or low-pass by selecting HP or LP outputs.)

† must return to ground via low-impedance dc path.

Fig. 7-9. Tuning the gain of a Q state-variable bandpass circuit.

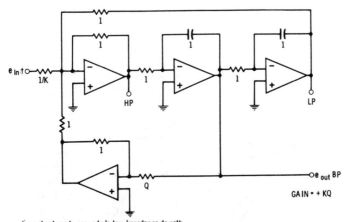

(A) Normalized to 1 ohm and 1 radian/sec.

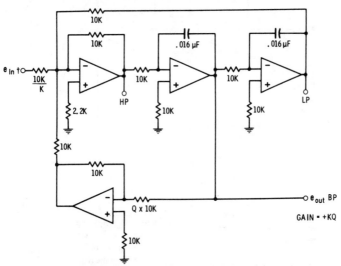

(B) Normalized to 10K and 1-kHz cutoff frequency.

Fig. 7-10. Variable-gain, state-variable bandpass circuit.

two integrators and an inverter. Loss is introduced into one of the integrators with a damping and Q-setting resistor. There is no high-pass output, and a possible low-pass output is only of very limited use.

The gain of the circuit is −Q for a unity input resistor and can be anything else, depending on the value of this resistor. The biquad is

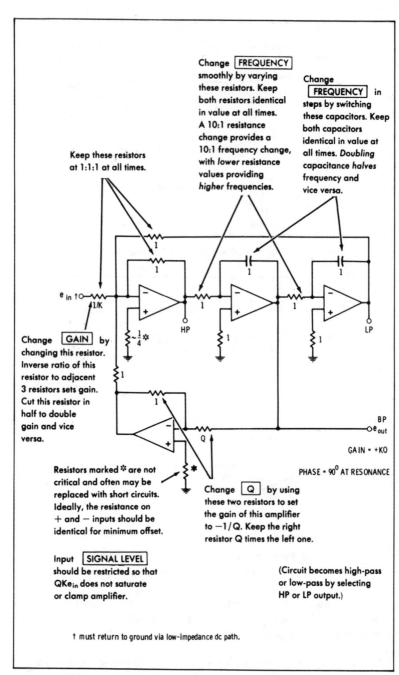

Change [FREQUENCY] smoothly by varying these resistors. Keep both resistors identical in value at all times. A 10:1 resistance change provides a 10:1 frequency change, with *lower* resistance values providing *higher* frequencies.

Change [FREQUENCY] in steps by switching these capacitors. Keep both capacitors identical in value at all times. *Doubling* capacitance *halves* frequency and vice versa.

Keep these resistors at 1:1:1 at all times.

e_{in}

1/K

Change [GAIN] by changing this resistor. Inverse ratio of this resistor to adjacent 3 resistors sets gain. Cut this resistor in half to double gain and vice versa.

$\sim \frac{1}{4}$ ✲

HP

LP

BP
e_{out}

GAIN = +KQ

PHASE = 90° AT RESONANCE

Resistors marked ✲ are not critical and often may be replaced with short circuits. Ideally, the resistance on + and − inputs should be identical for minimum offset.

Change [Q] by using these two resistors to set the gain of this amplifier to −1/Q. Keep the right resistor Q times the left one.

Input [SIGNAL LEVEL] should be restricted so that QKe_{in} does not saturate or clamp amplifier.

(Circuit becomes high-pass or low-pass by selecting HP or LP output.)

† must return to ground via low-impedance dc path.

Fig. 7-11. Tuning the variable-gain, state-variable bandpass circuit.

162

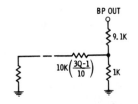

BP OUT

$10K\left(\dfrac{3Q-1}{10}\right)$

9.1K

1K

(A) Three-amplifier, unity-gain circuit.

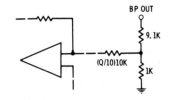

BP OUT

(Q/10)10K

9.1K

1K

(B) Four-amplifier, variable-gain circuit.

Fig. 7-12. Lowering the value of the Q-setting resistor.

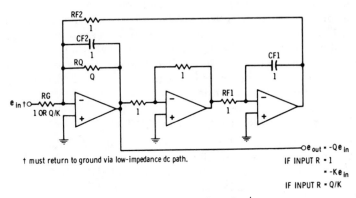

e_{in} ○

RG
1 OR Q/K

† must return to ground via low-impedance dc path.

$e_{out} = -Qe_{in}$
IF INPUT R = 1
= $-Ke_{in}$
IF INPUT R = Q/K

(A) Normalized to 1 ohm and 1 radian/sec.

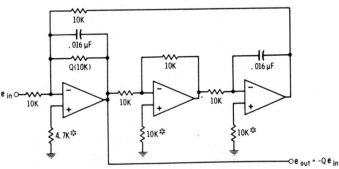

e_{in} ○

10K

$e_{out} = -Qe_{in}$

※offset resistors may be replaced with shorts in noncritical circuits.

(B) Normalized to 10K and 1-kHz cutoff frequency.

$$\frac{e_{out}}{e_{in}} = -\frac{RQ}{RG}\left[\frac{\dfrac{1}{(RQ)(CF2)}S}{S^2 + \dfrac{1}{(RQ)(CF2)}S + \dfrac{1}{(RF1)(RF2)(CF1)(CF2)}}\right]$$

(C) Transfer function.

Fig. 7-13. Biquad bandpass circuit (see text).

163

tuned by varying the capacitors in steps or by varying the tuning resistors. If we like, we can use a single-resistor tuning of the biquad, varying only RF2. The frequency will then vary as the square root of the resistor variation, giving us only a 3:1 or so variation for a 10:1 resistance change.

In contrast to the state-variable circuit, in the biquad circuit as the frequency changes, the bandwidth remains constant. This is the absolute bandwidth, not the percentage bandwidth. *The Q goes up and the percentage bandwidth goes down as frequency is increased, and vice versa.* If we used the low-pass output, we would find a very strong interaction between damping and center frequency.

The biquad is handy if you want a group of identical absolute-bandwidth channels in a system. This need is common in telephone applications, but otherwise it is rare. Generally, when you have a batch of channels, you want the higher-frequency ones to be proportionately wider than the lower-frequency ones, particularly in audio equalizers and electronic music.

FREQUENCY LIMITATIONS

The choice of op amp limits the frequency and Q that can be obtained, much more so with single IC circuits than with multiples. Fig. 7-14 gives some approximate recommended limits for the circuits of this chapter and the op amps of Chapter 2.

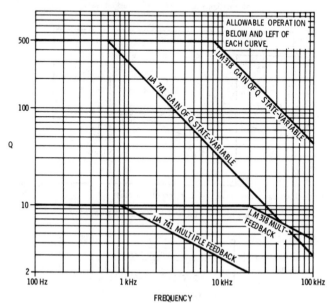

Fig. 7-14. Q and frequency limits for active bandpass filters, small output swings.

Note that this figure does not take large-signal slew rates into account. This may further restrict the maximum operating frequency if you need high-level outputs.

The tolerance and accuracy needed for cascaded filter sections were outlined in Chapter 5, particularly in Figs. 5-6, 5-12, 5-22, and 5-28.

SOME RULES

Any bandpass filter can be designed by using these guidelines:

1. If the percentage bandwidth is greater than 80 to 100 percent, use overlapping high-pass and low-pass filters instead. If the bandwidth is lower, use the circuits of this chapter.
2. Referring to the original problem and using Chapter 5, decide how many poles are needed and what their center frequency, Q, and staggering "a" are to be, and estimate their tolerances.
3. Pick a filter circuit, using the 1-kHz and 10K normalized values. The multiple-feedback circuit is recommended for fixed frequency, fixed low-Q (2 to 5) uses; the state-variable circuit for just about everything else.
4. Substitute the Q values needed for each stage.
5. Shift the stages by "a," multiplying or dividing the frequency-determining resistors by "a" as needed (fourth- and sixth-order filters only).
6. Scale the filter sections to their final center frequency, changing capacitor values by using Fig. 6-21 or by calculating the inverse frequency ratio.
7. Build, tune, and test the circuit.

Some examples are shown in Fig. 7-15. Most of the possible pitfalls of Figs. 6-23 and 8-22 also apply to the bandpass case. It is particularly important to maintain the component ratios at a constant value. Failure to do this can cause greatly different responses and strong interactions between tuning, Q, and gain.

RINGING, ELECTRONIC BELLS, AND THE TIME DOMAIN

Hit a pendulum with an impulsive hammer blow and it will oscillate for quite some time, dying out only when its nonperfect Q finally removes all the energy of the impulse. The same is true of electronic bandpass circuits. The narrower the bandwidth or the higher the Q, the longer the decay period, or time it takes to die out. We call this ringing the *transient response* of the pole.

Designing active bandpass filter circuits—some

A. Build a 3-kHz, Q = 30 bandpass pole.

We use the three-amplifier, state-variable circuit. From Fig. 7-14, we see that a 741 can handle the response. The Q resistor is calculated as 5K × (3Q — 1) or 440K. The capacitors will be 1/3 of .016 μF or 5600 pF. Final circuit looks like this:

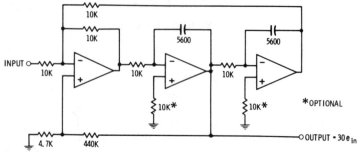

B. An octave-wide two-pole filter is to cover 200 Hz to 400 Hz with a 1-dB passband dip. Design the filter.

This is the same as the example in Fig. 5-18. From the previous results, we need a Q of 3.2, a staggering "a" value of 1.32, and a center frequency of 283 Hz. We can use a multiple-feedback section for each of the two poles. Resistance values initially will be 10K × 3.2 and 10K/3.2, or 32K and 3.12K. These must be lowered in value for the higher frequency pole and raised in value for the lower frequency pole by the "a" factor of 1.32. Final resistance values will then be 44K and 4.4K for the first stage and 24K and 2.4K for the second.

Frequency is scaled by increasing .016 μF by 1000/283 to .056 μF. The final circuit looks like this:

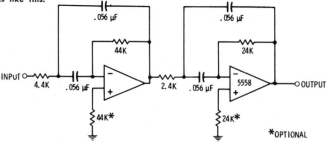

Had our example needed a sixth-order response, one of the poles is left at the center frequency and designed to one half of the Q of the other poles. The remaining two poles are raised and lowered in frequency by their "a" value. See Chapter 5 for examples.

Circuit gain will be 2Q², or approximately 16 dB per stage, for a total of 32 dB. From this the staggering loss of 11 dB must be subtracted, leaving us with a gain of 21 dB or slightly over 10:1.

Fig. 7-15.

> A single bandpass pole decays to $1/\epsilon$ of its initial amplitude in Q/π cycles
>
> $$(1/\epsilon \text{ is } 37\% \text{ of the initial value, or } 8.7 \text{ dB})$$

Fig. 7-16. Decay or "rundown" rule for a previously disturbed bandpass single pole.

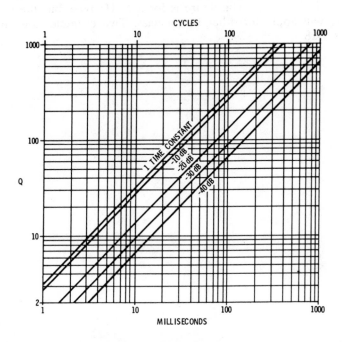

Fig. 7-17. Ringing or exponential decay of a 1-kHz pole.

Ringing and decay times in a bandpass pole—some **EXAMPLES**

A. A sudden transient is routed to a 1-kHz, Q = 50 pole. How long does it take for the oscillation to die below 30 dB of its initial value? How many cycles will this be?

From Fig. 7-17, we directly read a decay time of about 56 milliseconds, equivalent to 56 cycles as well.

B. Design an electronic C3 bell to drop to 10% amplitude in 0.8 second.

C3 is a frequency of 131 Hz (Fig. 10-3). An 800-millisecond ringing of this bell is the scaled equivalent of a ringing 131/1000 as long by a 1-kHz tone, or 105 milliseconds. From Fig. 7-17, we read the Q as approximately 140, noting that 10% amplitude is a 20-dB drop. The number of cycles to —20 dB will be approximately 105.

Fig. 7-18.

Sometimes, we might like to purposely use a bandpass filter as a ringing circuit, perhaps as an electronic bell, chime, or percussion instrument, or maybe in a quadrature art circuit (see Chapter 10). In these applications, we are more interested in the free-swinging, *time-domain* response than in the usual, forced *frequency-domain* response.

So, how long does it take a bandpass pole to die down, or decay? The rule is summed up in Fig. 7-16, and a useful plot of this rule appears in Fig. 7-17. The figure is for a 1-kHz pole, but the usual scaling will apply for other frequencies. Two examples are given in Fig. 7-18.

CHAPTER 8

High-Pass-Filter Circuits

In this chapter, we will learn how to build active high-pass circuits of orders one through six. The catalog circuits of this chapter are very similar to the low-pass ones of Chapter 6 and are essentially "inside out" versions of the same circuits with frequency-determining capacitors and resistors interchanged.

SOME HIGH-PASS RESTRICTIONS

Fig. 8-1 summarizes two important restrictions on the use of high-pass filters. In reality, there is no such thing as an active high-pass filter, for the upper-frequency rolloff of the operational amplifiers in use will combine to give a passband. If an active filter is to be good for anything, we have to save enough "daylight" between the lower passband limit of the active circuit and the upper limit set by the op amp. Very often, the maximum useful frequency limit for an active high-pass filter is a lot less than for an equivalent low-pass.

---

1. There is no such thing as an active high-pass filter. Op-amp response sets an upper frequency limit defining a response passband.

2. If a high-pass filter circuit also has a low-pass output inherently available, a dc bias path to ground *must* be provided at the input.

Fig. 8-1. Restrictions on an active high-pass filter.

If we have a *high-pass-only* type of circuit, the op-amp biasing paths are usually internal, and we do not have to worry any more about providing a dc return path through the input. This is not true if we are using a universal filter that consists of integrators or has high-pass, bandpass, and low-pass outputs, and we still have to provide the return path to ground through the input. Very often, operational-amplifer bias and offset problems are much less severe in high-pass circuits.

Another limitation of active high-pass circuits is that they are inherently noisier than low-pass ones. There are several obvious reasons for this. A high-pass filter is a differentiator that responds to sudden changes in inputs and uses this *transient* information to provide an output. In a high-pass filter, noise above the range of useful signals gets passed on, as do harmonics of "rejected" waveforms, unlike in a low-pass filter where these are strongly attenuated. Finally, in some op-amp circuits, the internal high-frequency limitations tend to decrease stability as frequencies near the upper limits are reached. The opposite is often true in many low-pass circuits where the op amp increases the damping and adds rolloff to the response.

FIRST-ORDER HIGH-PASS CIRCUITS

The first-order high-pass sections appear in Fig. 8-2. They are shown normalized, as usual, to one radian per second and one ohm and to 10K and a 1-kHz cutoff frequency. In the first-order section, the op amp is simply a voltage follower that frees the RC section from output loading.

The feedback resistor from the output to the inverting input is not particularly critical and should equal the value of the input-frequency resistor for minimum offset. As this frequency resistor will often be adjusted to adjust the frequency of this section, the feedback-resistor value often has to be a compromise that should be

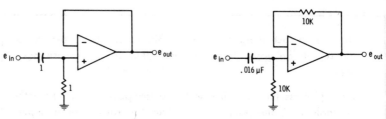

(A) Normalized to 1 ohm and 1 radian/sec.   (B) Normalized to 10K and 1-kHz cutoff frequency.

Fig. 8-2. First-order high-pass sections.

about the same as the average value of the tuning resistor. Often a blocking capacitor can be provided at the output of an active high-pass section to completely eliminate any offset shifts from appearing in the output.

SECOND-ORDER HIGH-PASS CIRCUITS

We can use the second-order low-pass circuits over again, rearranging them only slightly to get high-pass responses. The four low-pass responses were the unity-gain Sallen-Key, the equal-component-value Sallen-Key, the unity-gain state-variable, and the variable-gain state-variable.

Unity-Gain Sallen-Key Circuit

The math behind both Sallen-Key high-pass circuits appears in Fig. 8-3. As with the low-pass circuits, we have two cascaded RC sections driving an operational amplifier. The op amp both unloads the circuit from any output and feeds back just the right amount of

---

THE MATH BEHIND   Sallen-Key, high-pass, second-order sections.

Sallen-Key, second-order, high-pass filters can usually be redrawn into a passive network with an active source that looks like this:

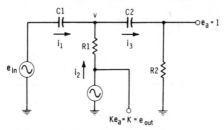

Since this network has to behave identically for any reasonable voltage at any point, it is convenient to force $e_a = 1$ volt and $e_{out} = Ke_a = K$. Solve for i_1, i_2, i_3 and then sum them:

$$i_3 = \frac{1 \text{ volt}}{R2} = \frac{1}{R2}$$

$$v = 1 + \frac{i_3}{j\omega C2} = 1 + \frac{1}{j\omega R2C2} = \frac{j\omega R2C2 + 1}{j\omega R2C2}$$

$$i_2 = \frac{K - v}{R1} \qquad i_1 = (e_{in} - v)j\omega C1$$

Fig. 8-3.

$$e_{in}(j\omega C1) = \frac{1}{R2} - \frac{K}{R1} + v\left(\frac{1}{R1} + j\omega C1\right)$$

$$e_{in}(j\omega C1) = \frac{j\omega R2C2}{\frac{1}{R2}(j\omega R2C2) - \frac{K}{R1}(j\omega R2C2) + (j\omega R2C2 + 1)(\frac{1}{R1} + j\omega C1)}$$

Rearranging and simplifying

$$\frac{1}{e_{in}} = \frac{(j\omega)^2}{(j\omega)^2 + \left[\frac{1}{R2C1} + \frac{1}{R2C2} + \frac{1-K}{R1C1}\right](j\omega) + \frac{1}{R1R2C1C2}}$$

And letting $S = j\omega$ and substituting for e

$$\boxed{\frac{e_{out}}{e_{in}} = \frac{KS^2}{S^2 + \left[\frac{1}{R2C1} + \frac{1}{R2C2} + \frac{1-K}{R1C1}\right]S + \frac{1}{R1R2C1C2}}}$$

Just as we did on the low-pass analysis, we have to restrict component values if K, frequency, and damping are to be independent. Two useful restrictions are:

(A) *Unity-gain, equal capacitors:* Let C1 = C2 and K = 1. For $\omega = 1$ R1R2C1C2 = 1, so C1 = C2 = 1 and R1 = 1/R2

$$\frac{1}{R2C1} + \frac{1}{R2C2} + \frac{1-K}{R1C1} = d = \frac{1}{R2} + \frac{1}{R2} = \frac{2}{R2}$$

$\therefore$ R2 = $\frac{2}{d}$ and R1 = $\frac{d}{2}$. The circuit looks like this:

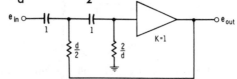

A unity-gain, equal-resistor realization is not possible for reasonable values of d.

(B) *All components identical:*
Let R1 = R2 = C1 = C2 for $\omega = 1$ R1 = R2 = C1 = C2 = 1

$$\frac{1}{R2C1} + \frac{1}{R2C2} + \frac{1-K}{R1C1} = d = 1 + 1 + 1 - K = d$$

and K = 3 − d. Note that this is the *only* value of K that will work properly. The circuit looks like this:

Fig. 8-3—continued.

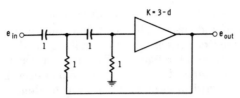

A circuit to provide a high input impedance and a gain of $3 - d$ is

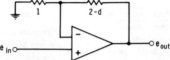

Gain is set by voltage-divider feedback, which is $\dfrac{1}{1+2-d} = \dfrac{1}{3-d}$, forcing the gain to $3 - d$.

Fig. 8-3—continued.

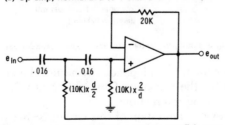

(A) Discrete, normalized to 1 ohm and 1 radian/sec.

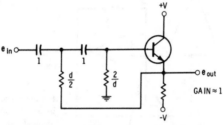

(B) Op amp, normalized to 1 ohm and 1 radian/sec.

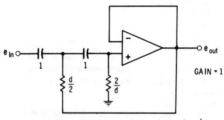

(C) Op amp, normalized to 10K and 1-kHz cutoff frequency.

Fig. 8-4. Simplest form of second-order high-pass active section—unity-gain Sallen-Key.

signal near the cutoff frequency to bolster the response to get the desired damping and shape. The main difference between the high-pass and low-pass circuits is that the positions of the resistors and capacitors have been interchanged. The circuit is shown in Fig. 8-4.

The main advantage of the unity-gain Sallen-Key is its extreme simplicity. In noncritical circuits it can even be done with a single-transistor emitter follower. The disadvantages are that the damping

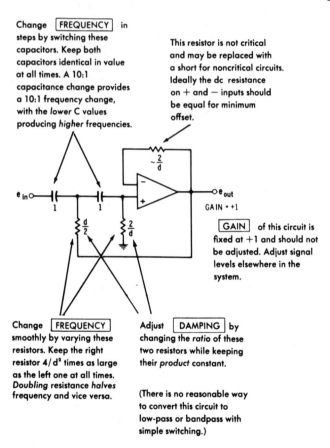

Change [FREQUENCY] in steps by switching these capacitors. Keep both capacitors identical in value at all times. A 10:1 capacitance change provides a 10:1 frequency change, with the *lower* C values producing *higher* frequencies.

This resistor is not critical and may be replaced with a short for noncritical circuits. Ideally the dc resistance on + and − inputs should be equal for minimum offset.

GAIN = +1

[GAIN] of this circuit is fixed at +1 and should not be adjusted. Adjust signal levels elsewhere in the system.

Change [FREQUENCY] smoothly by varying these resistors. Keep the right resistor $4/d^2$ times as large as the left one at all times. *Doubling* resistance *halves* frequency and vice versa.

Adjust [DAMPING] by changing the *ratio* of these two resistors while keeping their *product* constant.

(There is no reasonable way to convert this circuit to low-pass or bandpass with simple switching.)

Fig. 8-5. Adjusting or tuning the unity-gain, Sallen-Key, second-order high-pass section.

and frequency cannot be independently adjusted and that frequency variation calls for the tracking of two different-value resistors. Fig. 8-5 shows the tuning interactions.

Another inobvious limitation of this circuit is that you cannot simply interchange the components to turn it into an equivalent low-pass filter. Compare Fig. 8-4B with Fig. 6-5B. Note that the upper components are always in a 1:1 ratio and the lower are always in a $4/d^2$ ratio. The low-pass filter uses equal-value resistors; the high-pass filter uses equal-value capacitors. No simple switching of four parts can be used to interchange the two circuits.

Equal-Component-Value Sallen Key Circuit

The equal-component-value Sallen-Key circuit uses identical resistor values and identical values for capacitors. Thus, it is easy to switch the response from low-pass to high-pass and back again, provided we are willing to use a 4pdt switch per section. The circuits are shown in Fig. 8-6, and the tuning values and methods are shown in Fig. 8-7.

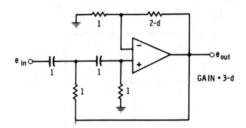

(A) Normalized to 1 ohm and 1 radian/sec.

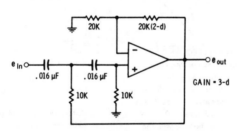

(B) Normalized to 10K and 1-kHz cutoff frequency.

Fig. 8-6. Equal-component-value, Sallen-Key, second-order high-pass filter has independently adjustable damping and frequency.

Like the low-pass circuits, these have a moderate positive gain. The damping is set by setting the gain. Damping and frequency may be independently adjusted. As usual, both capacitors must stay the same value and both frequency-determining resistors must remain identical in value at all times.

The ratio of the two resistors on the inverting input sets the gain and the damping. The absolute value of these resistors is not particularly critical. It is normally set so that the parallel combination equals the resistance seen from the noninverting input to ground. Note that

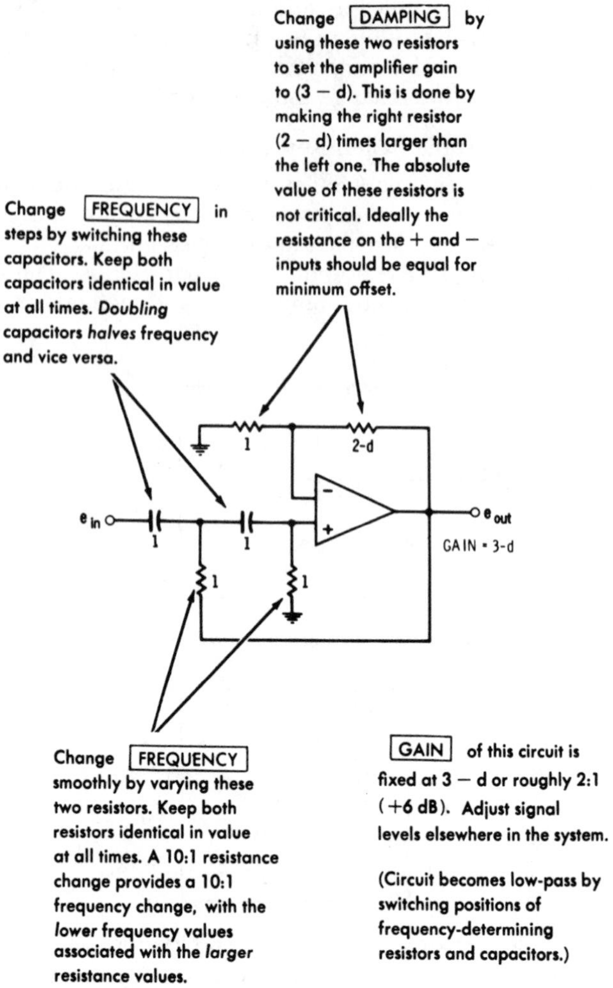

Change **DAMPING** by using these two resistors to set the amplifier gain to (3 − d). This is done by making the right resistor (2 − d) times larger than the left one. The absolute value of these resistors is not critical. Ideally the resistance on the + and − inputs should be equal for minimum offset.

Change **FREQUENCY** in steps by switching these capacitors. Keep both capacitors identical in value at all times. *Doubling* capacitors *halves* frequency and vice versa.

e in   GAIN = 3-d e out

Change **FREQUENCY** smoothly by varying these two resistors. Keep both resistors identical in value at all times. A 10:1 resistance change provides a 10:1 frequency change, with the *lower* frequency values associated with the *larger* resistance values.

GAIN of this circuit is fixed at 3 − d or roughly 2:1 (+6 dB). Adjust signal levels elsewhere in the system.

(Circuit becomes low-pass by switching positions of frequency-determining resistors and capacitors.)

Fig. 8-7. Adjusting or tuning the equal-component-value, Sallen-Key, second-order high-pass section.

the optimum-offset resistor values for low-pass will generally be *twice* that for the high-pass circuits, since we only have a single frequency-determining resistor returning directly to ground in the high-pass case. In a low-pass circuit, we have two frequency-determining resistors returning to ground through the source.

Offset is usually a much smaller problem in high-pass circuits. Usually, if your circuit is to switch between high-pass and low-pass, you use the optimum resistor values for the low-pass case.

Unity-Gain State-Variable Circuit

The math behind both state-variable filters is shown in Fig. 8-8, while the unity-gain circuit and its tuning appear in Figs. 8-9 and 8-10. We normally save the state-variable circuits for more elaborate

State-variable, high-pass, second-order sections.

An op-amp integrator looks like this:

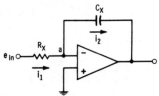

The high gain of the op amp continuously drives the difference between $+$ and $-$ input to zero. Point a is thus a virtual ground.

$$i_1 = \frac{e_{in}}{R_x} \text{ since point a is essentially at ground.}$$

$$i_2 = -\frac{e_{out}}{1/j\omega C_x} = i_1 = \frac{e_{in}}{R_x}$$

$$\frac{e_{out}}{e_{in}} = -\frac{1}{j\omega R_x C_x} \text{ or, letting } S = j\omega,$$

$$\frac{e_{out}}{e_{in}} = -\frac{1}{R_x C_x S}$$

The state-variable circuit

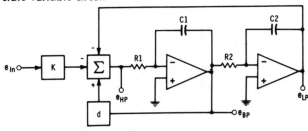

can now be analyzed:

$$e_{hp} = -Ke_{in} - e_{lp} + de_{bp} = e_{out}$$

$$e_{bp} = -\frac{e_{hp}}{SR1C1}$$

Fig. 8-8.

$$e_{lp} = -\frac{e_{bp}}{SR2C2} = +\frac{e_{hp}}{S^2R1C1R2C2}$$

$$Ke_{in} = -e_{hp} + de_{bp} - e_{lp}$$

$$(-K)e_{in} = e_{hp} + \frac{e_{hp}d}{SR1C1} + \frac{e_{hp}}{S^2R1R2C1C2}$$

which rearranges to

$$\frac{e_{out}}{e_{in}} = \frac{-KS^2}{S^2 + \dfrac{d}{R1C1}S + \dfrac{1}{R1R2C1C2}}$$

If $R1C1 = R2C2 = 1$, this becomes

$$\frac{e_{out}}{e_{in}} = \frac{-KS^2}{S^2 + dS + 1}$$

There are several ways to realize the summing block:

(A) Unity gain:

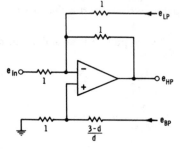

(B) Variable gain:

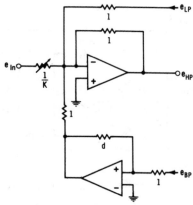

as previously analyzed in Fig. 6-9 and Chapter 2.

Fig. 8-8—continued.

or more critical jobs. They take three or four operational amplifiers per second-order section compared to the single one needed by the Sallen-Key circuits. State-variable circuits are often used where critical, low-damping values are needed, where voltage-controlled tuning is to take place over a wide range, where 90-degree quadrature outputs are needed, or where very simple switching between high-pass, bandpass, and low-pass responses is needed. They are also essential for the elliptical filters of the next chapter.

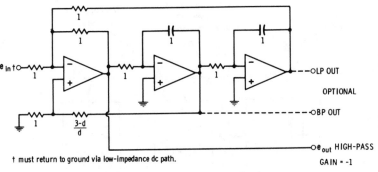

† must return to ground via low-impedance dc path.

(A) Normalized to 1 ohm and 1 radian/sec.

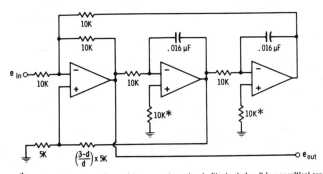

*optional offset compensation resistors - may be replaced with short circuit in noncritical applications.

(B) Normalized to 10K and 1-kHz cutoff frequency.

Fig. 8-9. Three-amplifier, state-variable filter offers unity gain, easy tuning, and easy conversion to low-pass or bandpass.

Since the circuit also provides low-pass and bandpass responses, a dc return path through the source must still be provided. Resistors on the noninverting inputs are optimized for minimum offset just as they were for the low-pass versions. The gain of the circuit is unity with a phase reversal.

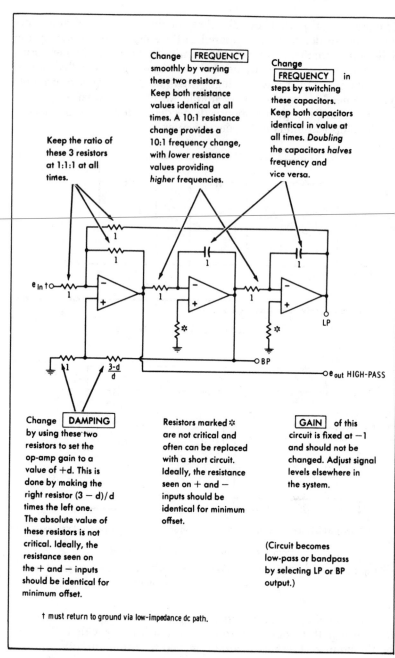

Change FREQUENCY smoothly by varying these two resistors. Keep both resistance values identical at all times. A 10:1 resistance change provides a 10:1 frequency change, with *lower* resistance values providing *higher* frequencies.

Change FREQUENCY in steps by switching these capacitors. Keep both capacitors identical in value at all times. *Doubling* the capacitors *halves* frequency and vice versa.

Keep the ratio of these 3 resistors at 1:1:1 at all times.

e_{In} †

LP

BP

e_{out} HIGH-PASS

Change DAMPING by using these two resistors to set the op-amp gain to a value of +d. This is done by making the right resistor $(3 - d)/d$ times the left one. The absolute value of these resistors is not critical. Ideally, the resistance seen on the + and − inputs should be identical for minimum offset.

Resistors marked ✻ are not critical and often can be replaced with a short circuit. Ideally, the resistance seen on + and − inputs should be identical for minimum offset.

GAIN of this circuit is fixed at −1 and should not be changed. Adjust signal levels elsewhere in the system.

(Circuit becomes low-pass or bandpass by selecting LP or BP output.)

† must return to ground via low-impedance dc path.

Fig. 8-10. Adjusting or tuning the unity-gain, state-variable, second-order high-pass section.

Variable-Gain State-Variable Circuit

If you want a *fixed* gain different from unity, you can recalculate resistor values for this gain. Since gain changes interact with the damping, a variable gain is not reasonable with the circuit of Fig. 8-9. The simplest way around this is to use a fourth op amp as shown in Fig. 8-11 and tuned as in Fig. 8-12.

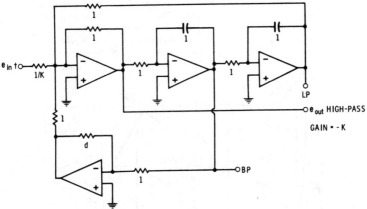

† must return to ground via low-impedance dc path.

(A) Normalized to 1 ohm and 1 radian/sec.

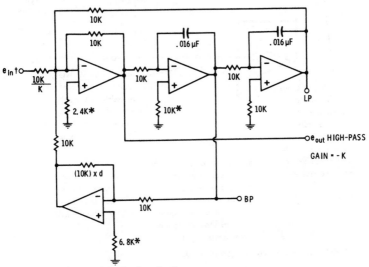

† must return to ground via low-impedance dc path.
*optional offset compensation resistors - may be replaced with short circuit in noncritical applications.

(B) Normalized to 10K and 1-kHz cutoff frequency.

Fig. 8-11. Variable-gain, state-variable filter.

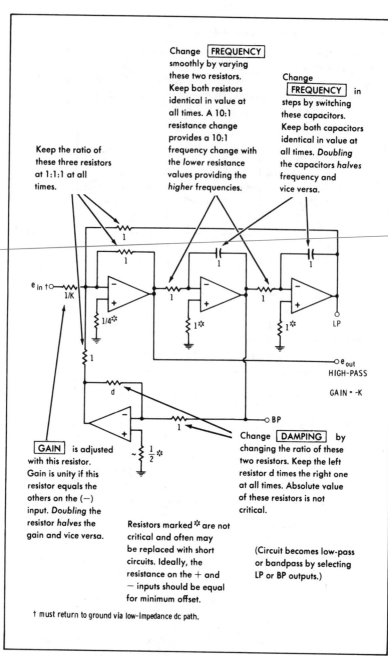

Change **FREQUENCY** smoothly by varying these two resistors. Keep both resistors identical in value at all times. A 10:1 resistance change provides a 10:1 frequency change with the *lower* resistance values providing the *higher* frequencies.

Change **FREQUENCY** in steps by switching these capacitors. Keep both capacitors identical in value at all times. *Doubling* the capacitors *halves* frequency and vice versa.

Keep the ratio of these three resistors at 1:1:1 at all times.

GAIN is adjusted with this resistor. Gain is unity if this resistor equals the others on the (−) input. *Doubling* the resistor *halves* the gain and vice versa.

Change **DAMPING** by changing the ratio of these two resistors. Keep the left resistor d times the right one at all times. Absolute value of these resistors is not critical.

Resistors marked ✳ are not critical and often may be replaced with short circuits. Ideally, the resistance on the + and − inputs should be equal for minimum offset.

(Circuit becomes low-pass or bandpass by selecting LP or BP outputs.)

† must return to ground via low-impedance dc path.

Fig. 8-12. Adjusting or tuning the variable-gain, state-variable, second-order high-pass section.

The gain is inversely set by the input resistor, independent of the damping. Damping, gain, and frequency are completely and independently adjustable. The output usually has a 180-degree phase inversion, which gives a stability advantage for higher gain values.

ANOTHER "RIPOFF" DEPARTMENT

Stock, Ready-to-Use Active High-pass Filters

For most simple applications, the equal-component-value Sallen-Key filter is often the best choice. Just as we have done with the low-pass filters, we can generate a catalog of response shapes of orders one through six for the seven shape options. These appear in Figs. 8-13 through 8-19.

Resistance values appear as 1% values, although the actual component tolerances needed for most of the circuits are typically 5% as shown. The damping-resistance values have remained the same as in the low-pass case, although their theoretically optimum values with regard to offset are one-half the values shown in these figures.

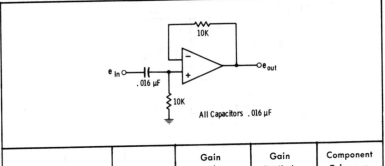

Response	RF1	Gain e_{out}/e_{in}	Gain Decibels	Component Tolerance
Highly Damped	10K	1	0	10%
Compromise	10K	1	0	10%
Flattest Amp	10K	1	0	10%
Slight Dips	10K	1	0	10%
1-Decibel Dip	10K	1	0	10%
2-Decibel Dip	10K	1	0	10%
3-Decibel Dip	10K	1	0	10%

To change frequency, scale all capacitors suitably. *Tripling* the capacity cuts frequency by *one-third,* and vice versa.

Fig. 8-13. First-order high-pass circuits, +6 dB/octave rolloff, 1-kHz cutoff frequency.

Optimum offset values will vary as the frequency-determining resistors are changed during tuning.

A third-order response can be approximated by a single operational amplifier as shown in Fig. 8-16. This is done by lowering the input impedance on the input RC section to one-tenth its normal impedance. This lowers the input impedance but also isolates any loading effects of the active section.

Since high-pass filters tend to be used with lower cutoff frequencies, scaling of the resistors to 100K or even higher can be done to lower the capacitor values. Offset problems will, of course, increase, but offset is rarely a problem in high-pass-only circuits until it gets so large it cuts into dynamic range or becomes temperature dependent or something equally drastic.

Capacitor values are all identical and are shown for 1 kHz. To scale capacitors to other cutoff frequencies, just calculate their inverse ratio or read the values from Fig. 8-20, a repeat of the curves of Fig. 6-21.

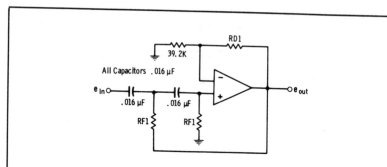

Response	RF1	RD1	Gain e_{out}/e_{in}	Gain Decibels	Component Tolerance
Highly Damped	12.7K	10.5K	1.3	2.3	10%
Compromise	11.3K	16.9K	1.4	3.0	10%
Flattest Amp	10.0K	22.6K	1.6	4.1	10%
Slight Dips	9.31K	30.9K	1.8	5.2	10%
1-Decibel Dip	8.66K	37.4K	2.0	6.0	10%
2-Decibel Dip	8.45K	43.2K	2.1	6.4	10%
3-Decibel Dip	8.45K	48.7K	2.2	6.8	5%

To change frequency, scale all capacitors suitably. *Tripling* the capacity cuts frequency by one-third, and vice versa.

Fig. 8-14. Second-order, high-pass circuits, +12 dB/octave rolloff, 1-kHz cutoff frequency.

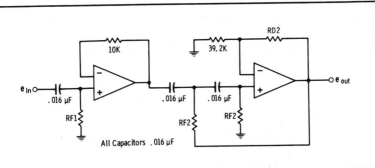

All Capacitors .016 μF

Response	RF1	RF2	RD2	Gain e_{out}/e_{in}	Gain Decibels	Component Tolerance
Highly Damped	13.3K	14.7K	21.5K	1.6	4.1	10%
Compromise	11.5K	12.1K	31.6K	1.8	5.1	10%
Flattest Amp	10.0K	10.0K	39.2K	2.0	6.0	10%
Slight Dips	6.65K	9.53K	51.1K	2.3	7.3	10%
1-Decibel Dip	4.53K	9.09K	59.0K	2.5	8.0	5%
2-Decibel Dip	3.24K	9.09K	63.4K	2.6	8.3	5%
3-Decibel Dip	3.01K	9.09K	66.5K	2.7	8.6	2%

To change frequency, scale all capacitors suitably. *Tripling* the capacity cuts frequency by *one-third*, and vice versa.

Fig. 8-15. Third-order high-pass circuits, +18 dB/octave rolloff, 1-kHz cutoff frequency.

The op-amp limitations rarely will interfere directly with the high-pass response. Instead of this, they usually place an upper limit on the passband. If we are to have a minimum of one-decade (10:1) frequency response well-defined as a passband, the limits of Fig. 8-21 are suggested for the op amps of Chapter 2.

As with the low-pass filters, there are lots of ways to get into trouble with these circuits. Common pitfalls are summarized in Fig. 8-22. Added to the low-pass restrictions are the generally worse noise performance you will get with a high-pass response and the need to save room for the passband between the filter and op-amp cutoff frequencies.

SOME HIGH-PASS DESIGN RULES

The following rules summarize how to use the circuits and curves of this chapter:

If you can use the equal-component-value Sallen-Key circuit:

1. Referring to your original filter problem and using Chapter 4, choose a shape and order that will do the job.
2. Select this circuit from Figs. 8-13 through 8-19 and substitute the proper resistance values.
3. Scale the circuit to your cutoff frequency, using Fig. 8-20 or calculating capacitor ratios inversely as frequency.

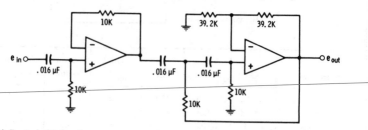

(A) Typical third-order, two op-amp filter (flattest amplitude, 1-kHz cutoff shown).

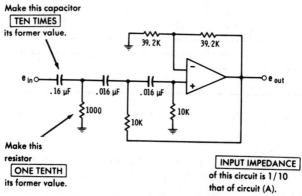

(B) One-op-amp approximation to (A).

Fig. 8-16. Approximating a third-order high-pass circuit with a single op amp.

4. Tune and adjust the circuit, using the guidelines in this chapter and Chapter 9. For very low frequencies, consider a 10X increase in impedance level to get by with smaller capacitors.

To build any active high-pass filter:

1. Referring to your original filter problem and using Chapter 4, choose a shape and order that will do the job, along with a list of the frequency and damping values for each section and an accuracy specification.
2. Pick a suitable second-order section from this chapter for each

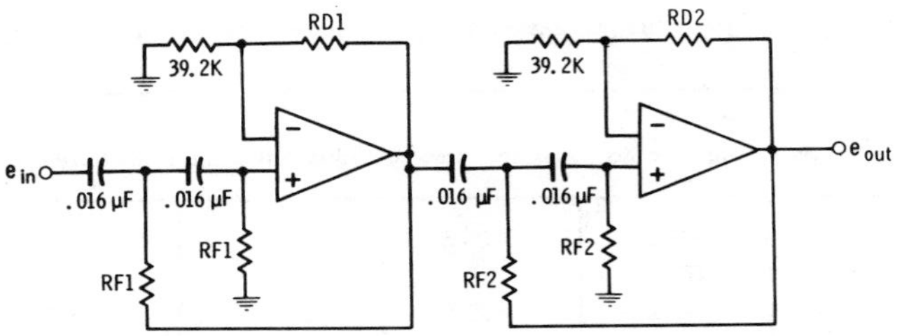

Response	RF1	RD1	RF2	RD2	Gain e_{out}/e_{in}	Gain Decibels	Component Tolerance
Highly Damped	14.3K	3.24K	16.2K	29.4K	+1.90	5.6	10%
Compromise	12.1K	4.64K	12.7K	41.2K	+2.30	7.2	10%
Flattest Amp	10.0K	5.90K	10.0K	48.7K	+2.60	8.3	5%
Slight Dips	7.15K	18.2K	9.76K	60.4K	+3.72	11.4	5%
1-Decibel Dip	5.23K	28.7K	9.53K	66.5K	+4.70	13.4	5%
2-Decibel Dip	4.64K	35.7K	9.53K	69.8K	+5.31	14.5	2%
3-Decibel Dip	4.42K	42.2K	9.53K	71.5K	+5.84	15.3	1%

To change frequency, scale all capacitors suitably. *Tripling the capacity cuts frequency by one-third, and vice versa.*

Fig. 8-17. Fourth-order high-pass circuits, +24 dB/octave rolloff, 1-kHz cutoff frequency.

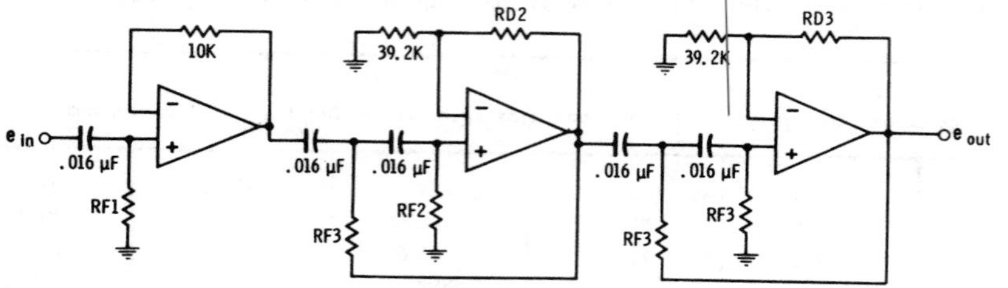

Response	RF1	RF2	RD2	RF3	RD3	Gain e_{out}/e_{in}	Gain Decibels	Component Tolerance
Highly Damped	15.4K	16.2K	8.82K	18.2K	35.7K	+2.3	7.2	10%
Compromise	12.4K	12.7K	12.1K	13.7K	46.4K	+2.8	9.0	10%
Flattest Amp	10.0K	10.0K	15.0K	10.0K	53.6K	+3.3	10.4	5%
Slight Dips	5.23K	8.06K	36.5K	9.76K	64.9K	+3.7	11.4	5%
1-Decibel Dip	2.80K	6.34K	49.9K	9.53K	71.5K	+4.7	13.5	2%
2-Decibel Dip	2.21K	6.19K	56.2K	9.53K	73.2K	+5.3	14.5	1%
3-Decibel Dip	1.78K	6.04K	59.0K	9.53K	73.2K	+5.9	15.4	1%

To change frequency, scale all capacitors suitably. *Tripling the capacity cuts frequency by one-third, and vice versa.*

Fig. 8-18. Fifth-order high-pass circuits, +30 dB/octave rolloff, 1-kHz cutoff frequency.

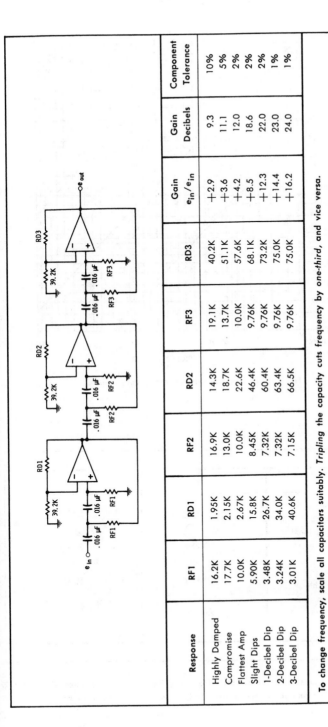

Response	RF1	RD1	RF2	RD2	RF3	RD3	Gain e_{in}/e_{in}	Gain Decibels	Component Tolerance
Highly Damped	16.2K	1.95K	16.9K	14.3K	19.1K	40.2K	+2.9	9.3	10%
Compromise	17.7K	2.15K	13.0K	18.7K	13.7K	51.1K	+3.6	11.1	5%
Flattest Amp	10.0K	2.67K	10.0K	22.6K	10.0K	57.6K	+4.2	12.0	2%
Slight Dips	5.90K	15.8K	8.45K	46.4K	9.76K	68.1K	+8.5	18.6	2%
1-Decibel Dip	3.48K	26.7K	7.32K	60.4K	9.76K	73.2K	+12.3	22.0	2%
2-Decibel Dip	3.24K	34.0K	7.32K	63.4K	9.76K	75.0K	+14.4	23.0	1%
3-Decibel Dip	3.01K	40.6K	7.15K	66.5K	9.76K	75.0K	+16.2	24.0	1%

To change frequency, scale all capacitors suitably. *Tripling the capacity cuts frequency by one-third, and vice versa.*

Fig. 8-19. Sixth-order high-pass circuits, +36 dB/octave rolloff, 1-kHz cutoff frequency.

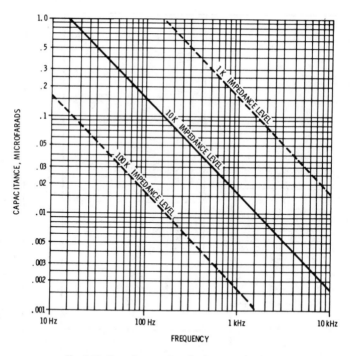

Fig. 8-20. Capacitance values for frequency scaling.

	741	LM318
Unity-Gain Sallen-Key	2.5 kHz	50 kHz
Equal-Component-Value Sallen-Key	1.0 kHz	20 kHz
Unity-Gain State-Variable	2.5 kHz	50 kHz
Gain-of-Ten (20 dB) State-Variable	250 Hz	5 kHz

Fig. 8-21. Recommended highest cutoff frequency limits for the op amps of Chapter 2. One decade minimum passband is provided with these limits.

1. Forgetting that high-pass circuits emphasize system noise, compared to low-pass ones that minimize it.

2. Forgetting to provide a low-impedance return path to ground on state-variable or other circuits that use integrators.

3. Damping resistors missing, wrong value, or too loose a tolerance.

4. R and C values not kept as specified ratios of each other.

5. Running an active high-pass cutoff frequency so close to op-amp high-frequency limitations that no passband remains.

6. Components too loose in tolerance or tracking poorly when adjusted. (See Chapter 9.)

7. Input signals too large, allowing saturation and ringing.

Fig. 8-22. Common pitfalls in high-pass active-filter circuits.

section you need, normalized to 1 kHz. Shift the frequencies of the cascaded sections as called for to realize the particular shape you are after. Remember that *increasing* resistance *decreases* frequency.

3. Set the damping value of each cascaded section to the value called for.

4. Scale the circuit to your cutoff frequency, using Fig. 8-20 or calculating capacitors inversely as frequency.

5. Arrange the circuits, starting with the highest-damped one near the input and lower-damped sections towards the output. Add a first-order active or passive section to the input if needed for an odd-order response.

6. Tune and adjust the circuit, using the guidelines of this chapter and Chapter 9. For very low frequencies, consider a 10X increase in impedance level to get by with smaller capacitors.

Several design examples of active high-pass filters appear in Fig. 8-23.

Designing active high-pass filters—some   EXAMPLES

A. Design a second-order rumble filter for a phonograph amplifier having a 1-dB peak and a 20-Hz cutoff frequency.

Use the circuit of Fig. 8-14, using 8.2K for the frequency resistors and a 39K re-

Fig. 8-23.

sistor for the damping resistors. Ten times the capacitor value scales to 100 Hz at 0.16 μF. Twenty Hz will be five times larger still, or 0.82 microfarads. While this capacitor is not particularly unreasonable, it is a bit expensive, which suggests raising the impedance of everything by a factor of 10. This gives us an input impedance of something around 100K for frequencies in the passband and gets us by with smaller 0.082 μF capacitors. The circuit looks like this:

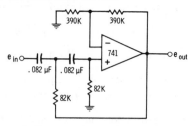

B. Design an electronic music filter to make the third harmonic of a C3 note (130.81 Hz) a minimum of 30 dB stronger than the fundamental. The note is a sawtooth.

The harmonic components of a sawtooth vary inversely in strength as the harmonic. Thus, the second harmonic is 1/2 the fundamental or 6 dB down, while the third harmonic is 1/3 the fundamental, or 10 dB down. Thus we start with the third harmonic —10 dB with respect to the fundamental and want to end up with the fundamental —30 dB with respect to the third. So, we apparently want to use a filter that will put the harmonic 40 dB above the fundamental.

If we use Fig. 4-1A, we see that a 3-dB-dips, high-pass filter peaks at 1.2 times its cutoff frequency. One-third this frequency is 0.4, the fundamental. We see the attenuation will be down by some 35 dB. Since a fourth-order filter would take a second op amp, we should probably try this one to see what it sounds like. So, we want a third-order, 3-dB-dips filter for trial. The cutoff frequency will be 130.8/0.4 = 327 Hz.

We use the circuit of Fig. 8-15, noting that frequency resistors are 3.01K for the first stage and 9.09K for the second stage, with a second-stage damping of 66.5K. This time, since capacitor values are not a problem but input impedance might be, we will leave the first-stage impedance where it was and scale the second by 10 to eliminate the op amp. The capacitor value is apparently .016/.327 = .047 μF for the first stage and 4700 pF for the second. The filter looks like this:

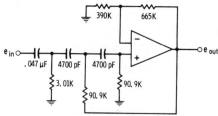

Two-percent tolerance components are recommended, but we could probably get away with 5% cut-and-try in this particular application.

C. Part of a complex synthesis problem calls for a section that simultaneously provides a 2.4-kHz low-pass and high-pass response and has a damping of 0.3. Design the section.

Fig. 8-23—continued.

This calls for a state-variable filter. We will assume that unity gain is acceptable. Capacitor values scale from $.016/2.4 = 6800$ pF and the damping resistor is $\left(\dfrac{3-d}{d}\right)5K = 9 \times 5K = 45K \approx 45.3K$. The final circuit is in this form:

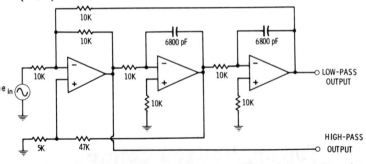

D. Show how we can convert the universal laboratory filters of Fig. 6-24D so that we can select either a low-pass or a high-pass response over the range of 10 Hz to 10 kHz.

For the Sallen-Key version, we arrange 8-pole double-throw switching to interchange the resistors and capacitors at the input to each op amp. Switching for a single section is as follows:

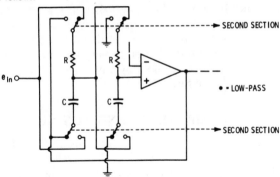

For the state-variable version, the switching is much simpler, but of course, six op amps are used instead of just two:

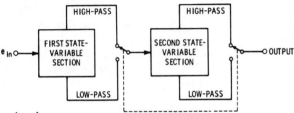

Fig. 8-23—continued.

CHAPTER 9

Tuning, Voltage Control, and Elliptical Filters

This chapter is about advanced techniques. We will look first at components and tuning methods, and then at ways to use electronic control, voltage control, or digital tuning with an active filter. From there, we will look into some notch and bandstop filters. We will end with an overview of *elliptical* filters, a very powerful, ultimate-falloff type of circuit that is really nothing but a very simple extension of the basic state-variable filter circuits.

COMPONENTS AND TUNING

It is always important to use the most accurate and most stable components you can get for any active filter system. One good choice of capacitor is the polystyrene type. It is stable and cheap and comes in many sizes to 5%, 2%, and even 1% tolerance. Its only disadvantages are that it melts if you touch it with a soldering iron and that some flux removers and solvents will attack it. So it must be used with reasonable care, but it is the all-around best bet for an active filter capacitor.

The second-best choices are Mylar capacitors for larger values and mica capacitors for smaller values. Mylar capacitors are limited in the choice of values and the tightness of the tolerance readily available. If you have a way to measure capacitance, parallel combinations can often be used to get a precision value at low cost.

NEVER use a disc ceramic or an electrolytic capacitor as a frequency component in an active filter. The discs are wide in tolerance and lossy, and their value changes with voltage and time. Electrolytics have the same disadvantages, in addition to being polarity sensitive and needing a bias voltage.

It should go without saying that you have to use good printed-circuit-layout techniques and reasonably high, well-regulated, properly bypassed power supplies. Circuit layout should obviously be as reasonably compact as possible, and the input and output portions of all active filters should be well separated from each other. Ground and supply systems should be broad foil and arranged so that no supply current or noise flows through the input ground-return connection.

Resistors

Ordinary carbon-film resistors are a good choice for many simpler active filters. Five-percent tolerance is usually good enough, although you can use parallel resistors to lower the value or series resistors to increase the value in final designs.

Newer molded metal-film resistors (Mepco/Electra) have 1% accuracy yet cost less than a dime in moderate quantities. They are thus far cheaper than traditional precision resistors.

Another obvious way to trim the value of a resistor is to put a small trimming pot in series with the resistor; this gives a 5% to 10% adjustment range. Several upright pc potentiometers can sometimes be controlled from a common bar shaft if they are placed just right on a pc board.

MANUAL WIDE-RANGE TUNING

The frequency of an active filter can usually be changed 10:1 by changing the capacitors to a value ten times or one-tenth their previous value. If we want continuous control, we can, in theory, use potentiometers. Normally, we need two pots per active section, which means four ganged pots for a fourth-order section and six for a sixth-order section. The problem is that reasonably priced snap-on multiple pots simply are not good enough for precision active filters, although they are just barely usable in economy circuits.

Fig. 9-1 shows a flattest-amplitude fourth-order filter we can adjust over a 10:1 range. Flattest-amplitude filters are usually the best choice for wide tuning since the value of the frequency resistor is the same in all sections, and the same is true of the capacitor value. Other responses require either a different resistor or a different capacitor value per cascaded section.

Ordinary quad pots, particularly the snap-together units, have some problems. The first is that resistance behavior is usually very unpredictable at the extremes of pot rotation, with the electrical rotation being much different and usually quite a bit shorter than

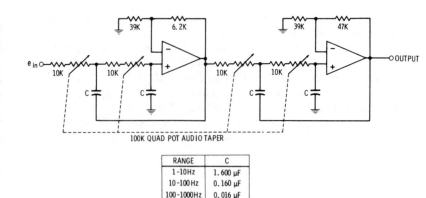

RANGE	C
1-10Hz	1.600 µF
10-100Hz	0.160 µF
100-1000Hz	0.016 µF
1-10kHz	1600 pF
10-100kHz	160 pF

Fig. 9-1. A 10:1 tunable fourth-order filter.

the mechanical rotation. A second problem is that we would like to see 5% tracking between the elements. This is much tighter than is normally provided, particularly with tapered or log pots. Both of these problems are solved with more expensive components, but the cost gets out of hand very quickly.

A third problem that we can do something about is the linearity. Since the frequency varies inversely with resistance, a linear pot will give a very cramped scale at one end. Fig. 9-2 shows how to beat this problem. In Fig. 9-2A, an ordinary linear pot takes half its rotation to get from a "1" to a "2" reading. In Fig. 9-2B, an ordinary audio log pot (10% cw log taper) makes things much worse. In Fig. 9-2C, a reverse taper (10% ccw log) straightens things out pretty well. The problem is, these are hard to find. Finally, in Fig. 9-2D, we use the same standard audio log pot that we used earlier; only, this time we put the *dial* on the pot *shaft* and the *pointer* on the *panel*. While the numbers go the usual way, the pot actually gets rotated backwards, reversing the taper for us. This last route is the simplest and cheapest way to get a reasonably linear scale.

One way to ease the pot problems is to restrict the tuning range, perhaps to 3:1 or less, and do more capacitor switching. A second is to go to digital or voltage control, which we will look at in just a bit.

A third route is popular with many commercial active-filter instruments. Instead of continuously adjusting the cutoff frequency, you use switched resistors. Fig. 9-3 shows one possibility. The cost is not much greater than when special pots are used, and the response is usually much more predictable and uniform. Enough switch positions are added to give the equivalent to continuous tuning. Usually the frequencies are arranged on a proportional log scale rather than

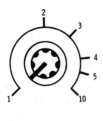

(A) Linear-taper pot.

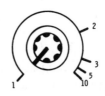

(B) Standard log-taper pot.

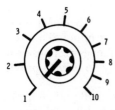

(C) Reverse log-taper pot.

(D) Standard log-taper pot. Dial on pot; pointer on panel.

Fig. 9-2. Using log pots and reverse dials to provide linear scales on tuning control.

on a linear basis. Ordinary 5% film resistors are usually adequate. Another advantage of switched control is that you can use different values for each section to build other response shapes and still use identical capacitors.

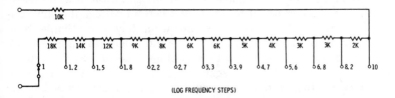

Fig. 9-3. Switched resistors give discrete cutoff frequencies, offer more precision, and afford better control.

Switching

We might like to use switches for changing from high-pass to low-pass and back again. With the state-variable filters, this is easily done with one spdt selector per second-order section or a dpdt switch for a fourth-order filter. A fourth-order Sallen-Key filter takes 8pdt switching, but it uses only 2 op amps instead of 6. One reasonably easy way to get 8pdt switching is to use a multiple-station 4pdt pushbutton switch. One button down puts the other button up, giving

8pdt switching, and the extra buttons can be used for off-on control and bypassing.

VOLTAGE AND DIGITAL CONTROL

Often, we would like to electronically change the frequency or Q of an active filter on a real-time basis, either by means of a control voltage or a digital computer command. This is particularly important in electronic music where a several hundred-to-one or higher frequency range might be desirable without mechanical switching. How do we go about it?

Fortunately, almost all of the circuits in this book let you change frequency widely without changing Q or damping. This is a major step in the right direction. If your filter frequency response cannot be independently controlled, there is essentially no way to provide voltage or digital control.

We expect two things of any voltage- or digital-control system. First, there must be just as many voltage-controlled "things" varying as there were pots in the original manual circuit. This is usually two resistors for frequency control per second-order section. We also have to force the "things" that are varying to be identical in value and within a certain tolerance, just as we did with the manual resistors.

Second, we can expect the inverse frequency relationship to hold—that is, if we raise the equivalent resistance of the "thing" we are using for tuning, the frequency goes down.

We can think of our frequency-determining resistor as a "network" that gives us a certain current out for a certain voltage in. This voltage-to-current relationship must be accurately controllable from unit to unit to guarantee reasonable tuning values. More important, if we replace the resistor with something else, it has to respond equally well to positive or negative signal swings. We say it has to be *bilateral*. In addition, we would like the "thing" replacing the resistor to be easy to drive, and we would not want any of the drive signal or any offset to appear in the output.

Electronic tuning is really quite simple. All we have to do is find a "thing" that acts like an electronically variable resistor. We can adjust the equivalent output current from a minimum possible value (minimum frequency) to a maximum possible value (maximum frequency) consistent with what the op amp can use and what it can drive—1000:1 should be theoretically possible, perhaps even more with newer, ultralow-input-current op amps.

So what is our "thing" going to be? Fig. 9-4 shows some older approaches to the electrically variable resistor problem. In Fig. 9-4A, we use a lamp and photoconductive cell. These used to be extremely nonlinear, but by replacing the lamp with a light-emitting diode, we can improve the linearity somewhat. In Fig. 9-4B, we use a junction field-effect transistor at zero bias and very low signal. This gives an electrically variable resistor, but it is not bilateral for large signal swings and it limits dynamic range and introduces distortion.

Fig. 9-4C uses a discrete MOS transistor with a floating substrate and feedback to linearize the response. This gives an electrically variable, bilateral resistor that handles up to several volts of signal with ease. We still get unit-to-unit variations, and transistors like the 2N4351 cost several dollars each. You can rework an ordinary hex inverter logic gate, the CMOS 4049, into six of these. They are useful at lower levels but will have distortion and dynamic range limitations.

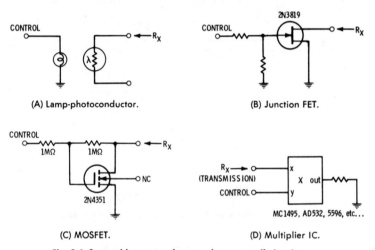

(A) Lamp-photoconductor.

(B) Junction FET.

(C) MOSFET.

(D) Multiplier IC.

Fig. 9-4. Some older approaches to voltage-controlled resistance.

Fig. 9-4D uses a "sledgehammer" technique. Each resistor is replaced with an integrated circuit called a four-quadrant multiplier. The multiplier in turn drives the lowest value of frequency resistor and scales the amplitude to make the output current of the resistor lower for lower input values. Very nicely, low input control voltages produce low output currents and low output frequency, and high input control voltages produce high output currents and high output frequency. Thus, we get a linear response to the cutoff frequency with respect to the input voltage. There are two restrictions to the circuit. First, you have to be sure the control voltage never gets below zero, or the output phase will reverse and latch the filter. Second, these units are expensive, particularly if you buy four or six at a time.

Fig. 9-5 shows how we can use electronic multipliers for voltage-controlled tuning of a state-variable filter section. Some suitable devices include the Motorola MC1495, the Signetics 5596, and the Analog Devices AD532. The connections and the external parts needed for trimming and bias vary with the device used, so always work directly with the appropriate data sheet and application notes.

Sections can be matched fairly accurately, and, by changing the output resistor slightly, we can accommodate different resistance values for cascaded sections of fancier response shapes.

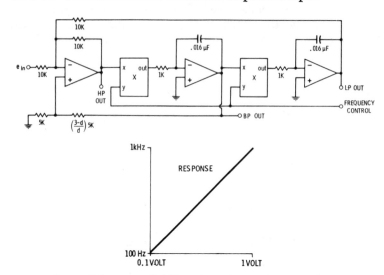

Fig. 9-5. Voltage-controlled filter using IC four-quadrant multipliers.

TWO NEW TECHNIQUES

Let us look at two relatively new integrated circuits, each of which gives us several reasonable and cheap ways to achieve acceptable accuracy with digital or voltage control. The first of these is the CMOS quad bilateral switch, the CD4016, available from most CMOS manufacturers (RCA, Motorola, Signetics, Fairchild, SSS, etc.). The second is the RCA CA3080, a low-cost variable-gain amplifier ideally suited to active filter use.

An Analog Switch

The CD4016 is shown in mini-catalog form in Fig. 9-6. It is simply a very good bilateral analog switch suitable for any off-on application. You apply supply power of +5 and −5 volts to the IC, and the analog input signal can be anything between these values up to 10 volts peak-to-peak. The control input is essentially an open circuit

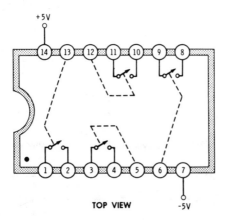

CMOS QUAD BILATERAL SWITCH

4016

+5V

TOP VIEW

-5V

This circuit contains four independent switches that may be used for off-on control of digital or analog signals. Signals to be controlled must be less than +5 and more than −5 volts.

+5 volts applied to pin 13 turns ON the connection between pins 1 and 2. −5 volts applied to pin 13 turns OFF the connection between pins 1 and 2. The other three switches are similarly controlled.

Input impedance to pin 13 is essentially an open circuit. The OFF resistance of pins 1 and 2 is many megohms; the ON resistance is 300 ohms. A lower-impedance, improved version is available as the 4066.

Fig. 9-6. A quad bilateral switch.

needing no drive current. A signal of +5 volts turns the switch ON and −5 volts turns it OFF. There are four completely independent switches in each package. Note that we use this device only as an OFF-ON controller. The on resistance is 300 ohms or so for the 4016 and 80 ohms for an improved version, the 4066.

Fig. 9-7 shows three ways to use the 4016. In Fig. 9-7A, we use it simply for high-pass/low-pass switching or for component insertion. The advantage of this method is that we can use dc control signals rather than having to route the actual active-filter signals through switches or other panel controls.

In Fig. 9-7B, we use four switches in combination to select combinations of resistors weighted 10K, 20K, 40K, and 80K. In combination, we get 16 different frequency values. The input is a 4-bit digital word. Obviously, we can use more switches and resistors for finer resolution.

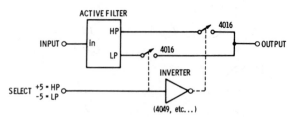

(A) Selecting outputs with dc control voltages.

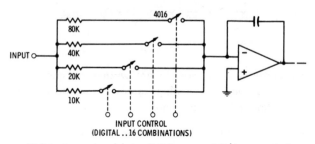

(B) Selecting resistors under digital command (D/A conversion).

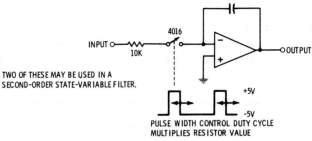

(C) Duty-cycle modulator provides variable resistance. Switching rate must be much faster than signal frequencies.

Fig. 9-7. Using the 4016 switch.

This process is called *digital-to-analog conversion,* and it permits direct control of the cutoff frequency of the filter with a digital word, derived either from local simple logic or from a computer. Note that this circuit is fully bilateral; many common d/a (digital-to-analog) converters are unilateral and will not work in this application.

Fig. 9-7C shows an interesting technique that is usable on low-frequency low-pass or bandpass state-variable filters. It changes the apparent value of fixed resistors by modulating their duty cycle. Suppose we place an analog switch in series with a frequency-determining resistor and turn it on and off at a very fast rate. The integrating capacitors of the filter will average out the rapid on-off input current fluctuations to some intermediate value.

Now, if we control the duty cycle or the percentage of the time the resistor is in the circuit, we can alter the filter cutoff frequency. A 10K resistor at a 50% duty cycle acts like a 20K resistor. At a 10% duty cycle, it acts like a 100K resistor, and so on, provided that the off-on switching is much higher in frequency than the signals and the time constants of the filter.

Variable-duty-cycle pulse generators are easy to build, particularly if you base them on a 555 timer, an 8038 function generator, an XR2240 astable, or any of many CMOS gate astable circuits. The beauty of this method is that we can get very accurate tracking for four, six, eight, or as many resistors as we like, and the price of each extra resistor is only one-fourth of a CD4016, or around 25¢. The limitations of the method are that the switching frequency must be much higher than the center or cutoff frequency and that noise, distortion, and feedthrough effects must be carefully controlled.

A TWO-QUADRANT MULTIPLIER

The RCA CA3080 is a transconductance amplifier that can be hooked up to make an ordinary output resistor act like an electronically variable one. The device is shown in Fig. 9-8. It looks somewhat like an op amp, but there are some important differences. The important thing is that we can use it as a linear voltage-versus-frequency gain-controlled amplifier. The cost of this circuit is under 50¢, in quantity. Fig. 9-9 shows its use in a state-variable filter.

This is a transconductance amplifier. It has a very *high* output impedance, provided by a bilateral current source. The output load resistor sets the voltage gain between input and output. In addition, a gain-control input varies the gain from the maximum possible set by the load resistor down to zero. As we provide more and more current to this pin, the gain of the circuit goes up proportionately. For a given input signal level, we get a variable and electronically controlled output current.

There are three important considerations when you use this device. The first is that input signals must be limited to 100 millivolts or less since the input to the circuit is an npn-transistor differential amplifier operating without feedback. Higher input signal swings will limit or clip. So, normally, you attenuate the active filter signals at the input and then build up with additional internal gain. The amount of gain will set the cutoff frequency of the filter.

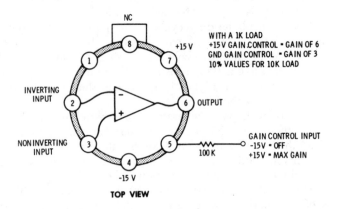

Fig. 9-8. A variable-gain amplifier that is a nearly ideal multiplier for active-filter work.

Second, remember that the output is a current, not a voltage, so the output resistor has a linear relationship with the voltage gain.

Third, the voltage-control input consists of the base of a transistor current, mirror connected to a negative supply. The input current sets the gain. *This input current must be limited by a series resistor, usually 100K or more.* Apply a voltage or ground this pin, and you destroy the IC.

The transconductance amplifier offers a very simple and low-cost way to provide linear control of the frequency of an active filter over

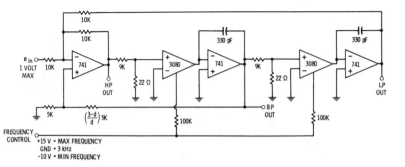

Fig. 9-9. Electrically variable filter using CA3080s.

a very wide range. The technique applies directly to the state-variable filter and others where one end of a resistor goes either to ground or to a virtual ground at the input of an op amp. With additional circuitry, it can be applied to "floating" resistors.

Log Operation

The CA3080 and the 4-quadrant multiplier provide a linear voltage-versus-frequency control. Sometimes, and particularly for electronic music, a log (exponential) characteristic is more desirable, so that a shift of one octave (2:1 frequency) always takes the same amount of control voltage, regardless of whether it is from 16 Hz to 32 Hz or from 2 kHz to 4 kHz.

We can build a log converter for the d/a conversion systems by providing a word changer called a *read-only memory* between the logic and the switches. For the analog circuits, the linear-to-log conversion is usually done by putting the base emitter junction of a transistor inside the feedback loop of an op amp. More details on this appear in various issues of *Electronotes*, in *The Nonlinear Circuits Handbook* by Analog Devices, and in *Operational Amplifiers* edited by Tobey, Graeme, and Huelsman (Burr-Brown Research Corp).

Note that all of the voltage- and digital-control techniques have been shown controlling the frequency of the filter. The same type of circuit can be used to control the Q or the damping. Most often, only a single device is needed per second-order section for Q setting, while two are needed for frequency setting. Remember that in composite, high-order filters, you have to proportion *all* filter sections *equally.* Thus, most sixth-order filters need six electronically variable resistors for frequency control and three of them for Q or damping.

NOTCH FILTERS

We can easily sum the outputs of high-pass, bandpass, and low-pass responses to get more-elaborate results. This is very easy to do

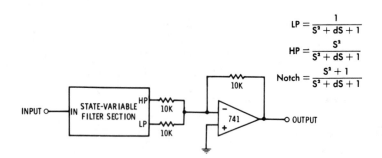

$$LP = \frac{1}{S^2 + dS + 1}$$

$$HP = \frac{S^2}{S^2 + dS + 1}$$

$$Notch = \frac{S^2 + 1}{S^2 + dS + 1}$$

Fig. 9-10. Notch filter built by summing low-pass and high-pass outputs of state-variable filter.

with the state-variable filter circuit that simultaneously gives us all three outputs. Two of the most useful responses are the notch, or bandstop, filter and the *elliptical,* or *Cauer,* response. Notch or bandstop filters are used whenever we want to reject or block a band of frequencies. Elliptical filters are similar to the low-pass and high-pass filters covered earlier in this book, except that they have the fastest possible falloff with frequency and a null or point of zero transmission just outside the passband.

Fig. 9-10 shows the basic notch configuration. We simply sum the high-pass and low-pass outputs of a state-variable filter and end up with a notch at the resonance frequency.

The response of a Q = 5 notch is shown is Fig. 9-11. The response of the notch and the bandpass are related. The resonance frequency of the bandpass response will equal the point of zero response of the notch. For reasonable Q values, the bandwidth of both responses will be the same. This means that the frequencies at which the bandpass response is down 3 dB from its peak resonance value will equal the frequencies at which the notch response also is 3 dB down. So, the

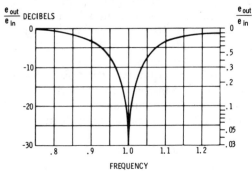

Fig. 9-11. Response of a Q = 5 notch filter.

bandpass of the bandpass response is equal to the response width of the notch, both defined to their 3-dB-down-from-peak-value frequencies.

As with the usual bandpass response curves, the notch response is symmetrical when you plot it on a log frequency scale.

If we did not want the notch to go to zero but only to some lower value, all we would have to do is sum part of the input signal with the high- and low-pass outputs of the filter. We can get any second-

---

THE MATH BEHIND **A transmission-zero filter.**

By changing one resistor value in Fig. 9-10, we can build an op-amp summing circuit that gives us a notch or transmission zero just outside a passband:

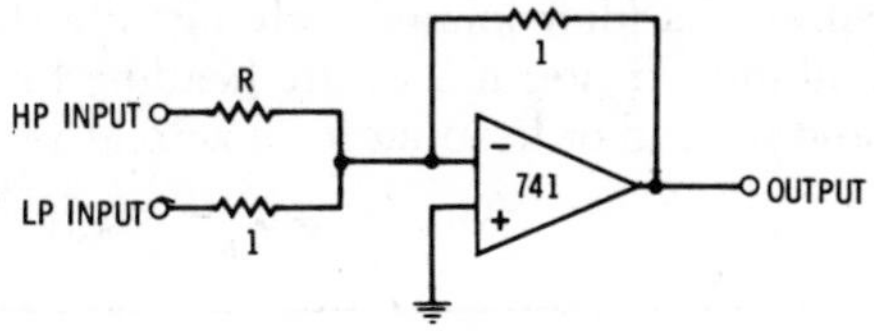

Assume we are using a low-pass response and $R > 1$

$$e_{out} = -\left[\frac{(e_{in})}{R}hp + \frac{(e_{in})}{1}lp\right]$$

If our HP signal is $\dfrac{S^2}{S^2 + dS + 1}$ and our LP signal is $\dfrac{1}{S^2 + dS + 1}$, the

sum will be $\left[\dfrac{S^2/R + 1}{S^2 + dS + 1}\right]$ or, letting $S = j\omega$, $\left[\dfrac{1 - \omega^2/R}{(1 - \omega^2) + jd\omega}\right]$

whose amplitude is

$$|e_{out}| = \frac{1 - \omega^2/R}{\sqrt{\omega^4 + (d^2 - 2)\omega^2 + 1}}$$

A transmission zero occurs at $1 - \omega^2/R = 0$ or at $\omega = \sqrt{R}$. At very high frequencies $\omega^4 >> \omega^2 >> 1$, so the response is approximately

$$\frac{\dfrac{\omega^2}{R}}{\sqrt{\omega^4}} = \frac{1}{R}$$

If $R = 1$, the circuit becomes the notch filter of Fig. 9-10.

Fig. 9-12.

order response we like simply by summing inputs and outputs from a state-variable circuit.

CAUER, OR ELLIPTICAL, FILTERS

Fig. 9-12 shows how we can change the gain of the high-pass output to produce a notchlike circuit that introduces a *transmission zero* just outside the passband of the filter. We can vary the frequency of the transmission zero with respect to the filter cutoff frequency by changing the ratio of high-pass to low-pass gain. Generally, the closer the gain is to unity, the sharper the response is, the closer the zero is to the filter cutoff frequency, and the more uniform the values of very high and very low frequency gain are.

This technique offers an obvious way to speed up the rate of falloff of the response of a filter immediately outside the passband, for things can fall off much faster if they are heading for a zero just outside the passband instead of heading for a zero at very high frequencies.

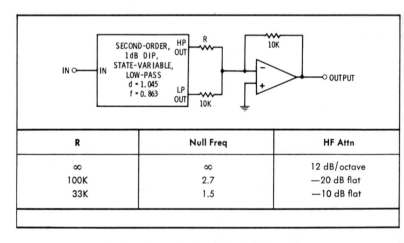

R	Null Freq	HF Attn
∞	∞	12 dB/octave
100K	2.7	—20 dB flat
33K	1.5	—10 dB flat

Fig. 9-13. Second-order, elliptical, low-pass filter.

If we design the filter to have the fastest possible falloff with frequency, we can use this transmission-zero technique to build a very strong class of filters called *Cauer*, or *elliptical*, filters. These filters fall off much more sharply outside the passband than the earlier filters described in this book. This makes them a very powerful design tool.

Of course, there has to be a catch. In exchange for this fast falloff, the filter response is allowed to bounce back up to some value well

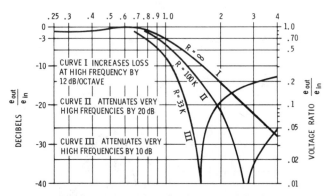

Fig. 9-14. Response of Fig. 9-13 circuit (second-order).

into the stopband and then either continue at a constant low value or else gradually fall off with increasing frequency. So, for frequencies *near* cutoff, you will get *more* attenuation, but for frequencies well into the stopband, you will get *less* attenuation than with the earlier filters.

Some typical examples appear in Figs. 9-13 through 9-18. The second-order response of Figs. 9-13 and 9-14 drops very fast with increasing frequency but bounces back up as shown. The third-order response curves do almost the same thing, only they continue falling off with frequency at a rate of 6 dB per octave above the bounce

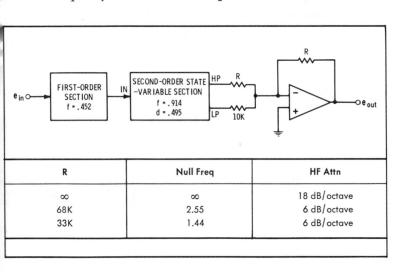

R	Null Freq	HF Attn
∞	∞	18 dB/octave
68K	2.55	6 dB/octave
33K	1.44	6 dB/octave

Fig. 9-15. Third-order, elliptical, low-pass filter.

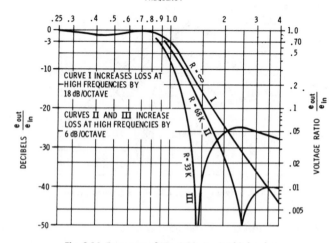

Fig. 9-16. Response of Fig. 9-15 circuit (third-order).

peak. Fourth-order curves are similar, only they end up falling off at 12 dB per octave above the bounce peak.

You can design filters like these just by starting with a conventional state-variable filter, perhaps the 1-dB-dips version, and then adding the single transmission-zero of Fig. 9-12. Varying the resistor on the high-pass output sets the notch frequency and the amount of bounce. While the math is rather complicated, it is a simple matter to change the one resistor and see what happens to the circuit.

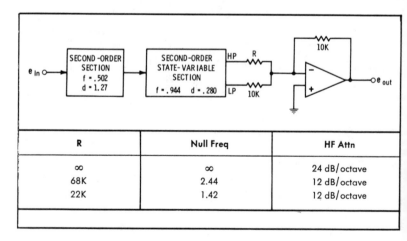

R	Null Freq	HF Attn
∞	∞	24 dB/octave
68K	2.44	12 dB/octave
22K	1.42	12 dB/octave

Fig. 9-17. Fourth-order, elliptical, low-pass filter.

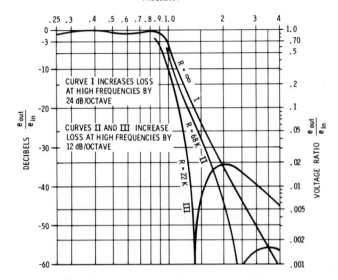

Fig. 9-18. Response of Fig. 9-17 circuit (fourth-order).

The transient and delay performance of these filters will be just about the same as the response of the filter you started with. Actually, for ideal passband response, the damping of one section should be lowered and its frequency should be raised, so an ideal elliptical filter will have somewhat poorer transient and delay response than an equivalent "all zeros at infinity" filter from earlier in the book.

We can generate equivalent high-pass filters simply by interchanging the inputs to the transmission-zero summer. However, the only type of filter that can be practically *switched* from high-pass to low-pass and back again is the flattest-amplitude, or *Butterworth*, response for orders of three and higher.

CHAPTER 10

Some Applications— What Good Are Active Filters?

So far, we have shown how to take the need for a particular filter and convert it into an actual working circuit. But what good are active filters? Where can we use them?

This chapter is about applications. It will show you many different things that can be done with active-filter techniques. Mostly, we will be zeroing in on uses that are simple and that provide new solutions to old problems. Rather than go into extreme detail, we will try to cover as many different application areas as we can. From there, you should be able to fill in the details yourself, using the earlier chapters and the references in this chapter as a design guide.

Many of the devices described in this chapter are available in kit form. One source is PAIA Electronics, Box 14359, Oklahoma City, OK 73114. A second source is Southwest Technical Products, 219 West Rhapsody, San Antonio, TX 78216.

BRAINWAVE RESEARCH

The human brain continuously generates low-frequency, low-amplitude electrical signals. With suitable scalp electrodes, these signals can be picked up, separated with active bandpass filters, and then monitored, displayed, or measured. Fig. 10-1 shows one possible system.

Systems of this type have clinical use in studies of epilepsy, stroke damage, schizophrenia, and related problems. More popularly, brainwave research leads to altered states of awareness, where visual, aural, or touch feedback of the brain's activity can lead to control of

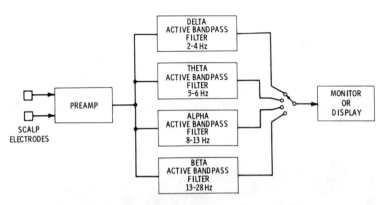

Fig. 10-1. Brain-wave (biofeedback) monitor using active filters.

emotions and creative drives as well as achieving transcendental meditative states similar to those of yoga, mind-altering drugs, or strict religious training. Positive feedback of the brain's activity has also been demonstrated to be useful in attaining partial control of normally automatic body functions. Relief or elimination of migraine headaches is one demonstrated possibility.

Four characteristic frequency bands are recognized. The lowest are the *delta* waves between 2 and 4 Hz; these are associated with deep sleep and young infants. *Theta* waves of 5 to 6 Hz are related to self control, annoyance, and frustation. *Alpha* waves range from 8 to 13 Hz and are associated with awareness and relaxation. Finally, *beta* frequencies of 13 to 28 Hz seem to be related to tension or surprise.

The normal signal level at a scalp contact is a few dozen microvolts at most, and quality brainwave instruments have to provide a sensitivity of a very few microvolts for adequate pickup. Active filters are ideal for such low frequencies, using the bandpass techniques of Chapter 7. Moderately to highly damped filter versions are recommended for minimum transient effects.

More information on these techniques appears in *Altered States of Awareness,* a book by T. J. Teyler. An alpha wave construction project appeared in the January 1973 *Popular Electronics.*

ELECTRONIC MUSIC

Active filters are such an integral part of today's electronic music scene that it is hard to imagine what it would be like without them. Let us look at four popular application areas—modifiers of conventional instrument sounds, formant filters, synthesizer vcf's (voltage-controlled filters), and percussion generators.

Modifiers

An active filter added to a guitar preamplifier can dramatically change the sound of the instrument by selectively emphasizing or de-emphasing portions of the acoustic spectrum of the guitar. A typical instrument is shown in Fig. 10-2, and technical details appeared in the June 1974 *Radio Electronics.* Similar active circuits can alter or modify the sound of virtually any conventional instrument.

Courtesy Southwest Technical Products Corp.

Fig. 10-2. Guitar preamp uses active filters as modifiers.

Formant Filtering

There are two fundamentally different ways of modifying electronic tone sources in electronic music. If we use a fixed-filter system, we are using *formant filtering.* The harmonics of each note of differing frequency vary in structure. This is also the case with many conventional musical instruments where their size and shape provide a fixed acoustical filtering response. On the other hand, if we use a *voltage-controlled filter,* or vcf, the harmonics of each note can be pretty much the same and usually independent of frequency. This creates a distinct "electronic" or "synthesizer" sound.

The frequencies of the various notes involved in Western music appear in Fig. 10-3. There are usually twelve notes per octave. An octave is a 2:1 frequency range, and the note frequencies repeat in higher octaves in a 1, 2, 4, 8, . . . sequence.

Fig. 10-4 shows some formant-filtering techniques that make optimum use of active filters. Start with a square wave and filter it lightly with a low-pass filter. You will get a group of tone structures that sound "hollow" or "woody," similar to the sound of such instruments as clarinets. Minor filter action on a sawtooth wave leads to string voices, made brighter by high-pass filtering or more mellow through low-pass emphasis. The same sawtooth routed through a bandpass

| Octave | Note | | | | | |
Number	C	C♯	D	D♯	E	F
0*	16.352	17.324	18.354	19.445	20.602	21.827
1	32.703	34.648	36.708	38.891	41.203	43.654
2	65.406	69.296	73.416	77.782	82.407	87.307
3	130.81	138.59	146.83	155.56	164.81	174.61
4	261.63	277.18	293.66	311.13	329.63	349.23
5	523.25	554.37	587.33	622.25	659.26	698.46
6	1046.5	1108.7	1174.7	1244.5	1318.5	1396.9
7	2093.0	2217.5	2349.3	2489.0	2637.0	2793.8
8	4186.0	4434.9	4698.6	4978.0	5274.0	5587.7

| Octave | Note | | | | | |
Number	F♯	G	G♯	A	A♯	B
0*	23.125	24.500	25.957	27.500	29.135	30.868
1	46.249	48.999	51.913	55.000	58.270	61.735
2	92.499	97.999	103.83	110.00	116.54	123.47
3	185.00	196.00	207.65	220.00	233.08	246.94
4	369.99	392.00	415.30	440.00	466.16	493.88
5	739.99	783.99	830.61	880.00	932.33	987.77
6	1480.0	1568.0	1661.2	1760.0	1864.7	1975.5
7	2960.0	3136.0	3322.4	3520.0	3729.3	3951.1
8	5919.9	6271.9	6644.9	7040.0	7458.6	7902.1

*Octave zero is very seldom used. Frequencies shown are in hertz and are valid for any electronic musical instrument, organs, and any conventional instrument except the piano.
"Middle C" is C4 at 261.63 Hz. Standard pitch is A4 = 440.

Fig. 10-3. Standard frequencies for the 12-note, equally tempered music scale.

filter produces horn sounds. Multiple-spectrum instruments such as the bassoon, English horn, and oboe take multiple-bandpass filters or a notch filter. Heavy filtering of a sawtooth will recover only the fundamental with slight second-harmonic components, characteristic of the flute and some organ voices.

Realistic imitation of traditional instruments depends both on the control of the harmonics and the envelope or amplitude of the note. Usually the envelope and formant filtering are done separately and then combined in a voltage-controlled amplifier (vca), a keyer, or a multiplier.

Voltage-Controlled Filters

The electronic tuning techniques of Chapter 9 let us build wide-range voltage-controlled filters well-suited for electronic music use. One important advantage of the vcf route is that you can change the harmonic structure of a note during its existence, particularly on the attack (rise time) and decay (dying out or fall time) portions. Volt-age-controlled filter techniques are more often associated with single-voiced or monophonic music systems, while formant-filtering systems are more common with polyphonic systems having many or overlapping notes simultaneously possible. Two synthesizers using voltage-controlled filter techniques appear in Figs. 10-5 and 10-6.

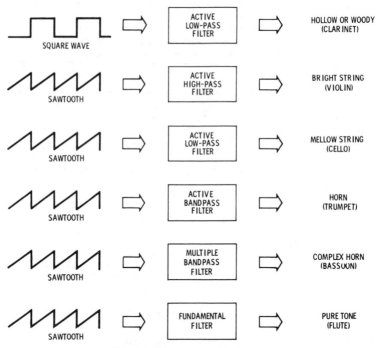

Fig. 10-4. Active formant filter basics.

Percussion Generators

The transient response of a bandpass filter can be used as a simple bell or chime by following the design guidelines of Chapter 7. With several transient generators, we can produce most of the percussion music voices, particularly if some white noise or modified noise is properly introduced at the same time. A two-step decay often gives the most realistic responses. A struck bell or cymbal has a distinct "clang" at the instant it is struck, followed by a long drawn out decay of relatively pure tone.

For more information on electronic music technique, see *Electro-notes* (203 Snyder Hill Road, Ithaca, NY 14850), a series of articles beginning in the September 1973 *Popular Electronics,* and the February 1974 *Radio Electronics.* Synthesizer techniques are often cov-

Fig. 10-5. Synthesizer using VCF techniques.

Fig. 10-6. "Gnome" microsynthesizer uses active voltage-controlled filter.

ered on a professional level in the *Journal of the Audio Engineering Society,* and the characteristic sounds of traditional instruments are often analyzed in depth in the *Journal of the Acoustical Society of America.*

AUDIO EQUALIZERS

Any time you want to emphasize or de-emphasize a portion of the audio spectrum, an active filter can be used. Most often, we are dealing with damped filters for low-pass and high-pass responses and with very moderate Q values (from 2 to 5) for bandpass applications

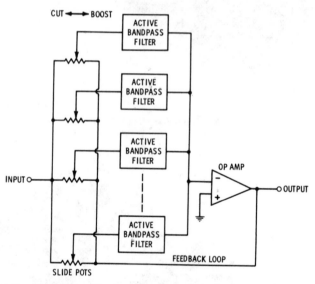

Fig. 10-7. Graphic equalizer uses multiple active filters within a single feedback loop.

for general audio use. This means that the active filters become extremely simple and economical to use for general audio applications.

One currently popular form of spectrum modifier is the *graphic equalizer.* This most often consists of a bank of slide potentiometers that emphasize or de-emphasize portions of the audio spectrum. The graphic equalizer is used to improve or match apparent room acoustics, to modify or match instruments, to add special effects to an otherwise dull recording, to improve speech intelligibility on a noisy channel, and similar tasks.

Often the individual filter channels are made of single-op-amp, low-Q, active bandpass filters. These are usually placed inside the feedback loop of an operational amplifier as shown in Fig. 10-7. This

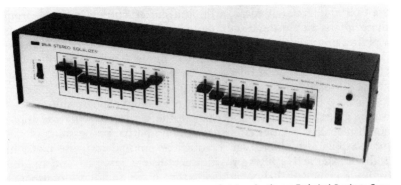

Courtesy Southwest Technical Products Corp.

Fig. 10-8. Graphic equalizer uses eighteen active filters.

feedback provides a "boost" or "cut" operation where each slide control provides a flat response in the middle of its range and provides progressively more emphasis or de-emphasis as the limits of the slider are approached. Fig. 10-8 shows a kit stereo 18-channel equalizer; Fig. 10-9 shows one of the internal circuit boards. Note the extreme simplicity and compactness of the circuit. More details on these equalizer techniques appear in the May 1974 *Popular Electronics*.

QUADRATURE ART

The state-variable active filter has opened up an interesting new art form we can call *quadrature art*. It is based on the fact that there

Courtesy Southwest Technical Products Corp.

Fig. 10-9. Inside of equalizer shows nine active filters. Note extreme simplicity and compactness.

are two 90-degree (quadrature) phase-shifted outputs available from these filters. Route these outputs to an X-Y display system, such as a plotter or an oscilloscope, and drive them from a source of interesting audio signals and you generate unique families of constantly changing closed-line art forms. The basic setup is shown in Fig. 10-10, and some of the simpler response curves appear in Fig. 10-11.

The key to the system is that two 90-degree phase-shifted sine waves generate a circle. The size of the circle is set by the input amplitude. If the input is changing, the relative phase shifts of the two outputs also change, generating the unusual response plots. In addition, the active filter provides a transient response and ringing to sudden changes that get added to the normal steady-state response to a slowly changing input. Together the two add up to a wild variety of possibilities.

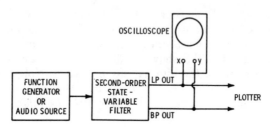

Fig. 10-10. Basic quadrature art setup.

It is rather hard to set down some rules for this new art form. Normally the frequencies of the input signals are much lower than the center frequency of the active filter. The filter Q sets the tightness or the number of turns of any spiral. The total number of lumps in the response depends on the ratio of input frequency to filter frequency. The amplitude of the input sets the size of the output display. Gradual input changes elicit the forced or steady-state filter response, while abrupt input changes produce the transient response.

Varying the ratio of the X gain and the Y gain changes the aspect ratio of the display, from elliptical to circular to elliptical forms again. The photos of Fig. 10-11 show fixed, single-input, steady-frequency responses to the various square, ramp, pulse, and triangle outputs of an ordinary function generator. You can get considerably more-elaborate displays when the inputs are changing or are a combination of many sources. Driving the filter from a music synthesizer offers many dynamic possibilities.

Raster-scan (tv-type) displays present a few problems and cannot directly be driven by a quadrature art system, except at very low speeds. One way around this is to remove the yoke from the picture tube of a conventional tv set, leaving it connected to the rest of the

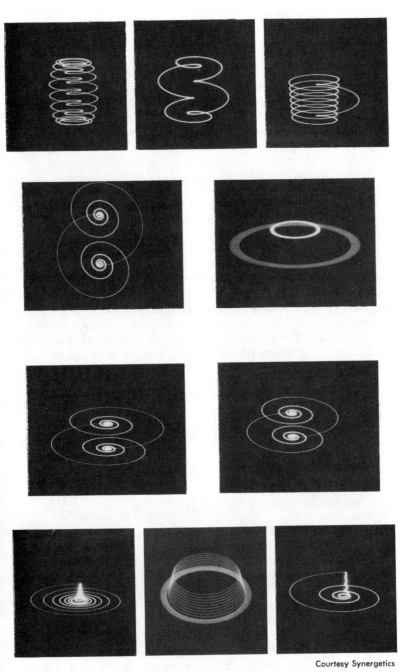

Courtesy Synergetics

Fig. 10-11. Some elementary quadrature art forms.

circuit. A new yoke is then placed on the picture tube and driven from a stereo amplifier. This converts the tv from a raster scan to an XY scanning system.

Some spot protection that blanks the display when the input amplitudes of both deflection amplifiers hit zero is recommended. This prevents phosphor burns and gives the best appearance. Another approach to raster-scan display of quadrature art is to use some form of digital memory between the active filter and the final raster-scan display.

OSCILLATORS AND SIGNAL SOURCES

Provide enough external feedback to almost any filter and you can convert the filter into an oscillator or signal source. Let us look at a typical example:

Suppose you wanted a constant-amplitude, low-distortion sine-wave oscillator that you could voltage-tune, say, over a 1000:1 frequency range or more. How would you go about it?

Most of the traditional ways to do this (heterodyning two radio frequencies, Wein bridge circuits with agc loops, diode break distortion networks, etc.) have one or more problems that limit the design or performance. How can active filters with feedback do any better?

Fig. 10-12 shows one possibility. We build a state-variable, electronically tunable (see Chapter 9) active bandpass filter with unity

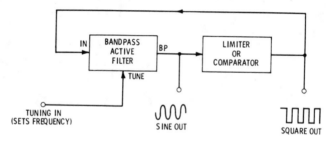

Fig. 10-12. Wide-range sine-wave/square-wave signal source using state-variable or biquad active filter.

gain, covering the frequency range of interest. We take the bandpass output and run it into a constant-amplitude square-wave converter, such as a limiter or a comparator. We then take the output square wave and route it right back to the input of the active filter.

What happens? The square wave is always of constant amplitude. Its fundamental frequency always equals the center frequency of the active filter (except during sudden changes), and the output amplitude of the sine wave will also be constant.

A square wave consists of a fundamental and a group of odd harmonics. The third harmonic will be 1/3, or 33%, of the amplitude of the fundamental. The fifth harmonic will be 1/5, or 20%, and so on. The bandpass filter is set to the fundamental frequency of the square wave, so it strongly attenuates the third harmonic and even more strongly attenuates the higher harmonics.

As you change frequency, the active filter temporarily adds or removes enough phase shift while the frequency is shifting to move the output in the right direction to reach the new frequency.

The Q of the active filter trades off the sine-wave distortion against the speed with which you can slew or sweep the response without transient effects during a frequency change. For instance, with a Q of 10, Fig. 5-6 tells us that we would get an attenuation of around 28 dB to the third harmonic. This would reduce the distortion to slightly over 2%. A distortion of 0.2% can theoretically be obtained with a Q of 40, again with values read from Fig. 5-6, and higher Qs will yield still lower distortion values.

On the other hand, with a Q of 10, you will not want to change frequency faster than something like 10% per cycle unless you want violent changes in amplitude during the change process. With a Q of 40, this limit moves down to 2.5% per cycle and so on. If you change frequency well under these limits, the amplitude stays constant as you change frequency. Above these limits, you can go through some wild amplitude gyrations before you get a final output of stable value. By the way, it is important to make sure the trip points on the sine-to-square-wave converter are identical for positive and negative cycles; otherwise, some second-harmonic distortion will be introduced.

As plus benefits of this technique, you get a "free" pair of 90-degree phase-shifted outputs if you use the low-pass output as well, along with the equivalent frequency square wave.

TEST AND LAB FILTERS

A universal, wide-range fourth-order active filter is useful for general-purpose audio testing, eliminating noise in experimental setups, designing tone systems, and so on. You can build one yourself, using the guidelines of Figs. 6-24 and 8-23. A version that gives fourth-order high- and low-pass responses, which can be switched in decade steps and adjusted smoothly over a 10:1 range, works well for many applications. Simple switching allows for either Bessell (highly damped) or Butterworth (maximum-flatness or critically damped) responses. One possible package is shown in Fig. 10-13.

These instruments are also available commercially, either as a complete system (Fig. 10-14) or as individual modular blocks (Fig.

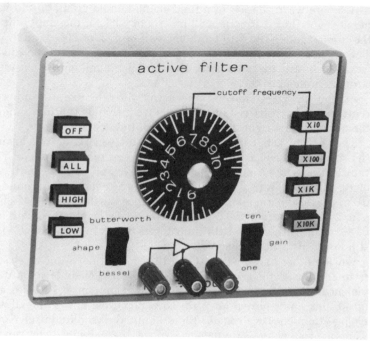

Fig. 10-13. General-purpose lab or shop filter.

Fig. 10-14. Precision variable active filter.

Courtesy Burr-Brown Research Corporation

Fig. 10-15. Commercial active-filter modules.

10-15). Bandpass filters as well as high/low versions are also possible. One configuration would have two poles and control of pole Q and staggering. A group of fixed-frequency active filters is shown in Fig. 10-16.

Courtesy Frequency Devices, Inc.

Fig. 10-16. Group of fixed-frequency active filters.

SPEECH THERAPY

Visual feedback of sound is useful as a training aid to help cure speech impediments such as stuttering and some forms of mental retardation. Fig. 10-17 shows one possibility. We take a microphone pickup and divide the audio into many narrow channels using active bandpass filters. The energy in each channel is detected and used to control either a single colored lamp or LED, or else to combine the outputs on some sort of bar graph or color-tv display. Similar techniques are involved in speech analysis and computer-based artificial-speech-generation circuitry.

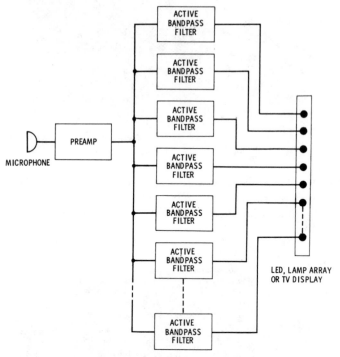

Fig. 10-17. Speech-therapy system uses active bandpass filters.

AN ALWAYS-ACCURATE CLOCK

One method of making a digital clock self-resetting and always accurate is to tie it into one of the time services of the National Bureau of Standards. These services present timing information that can be recovered as a parallel code. The code can then be loaded into a clock to correct its time display.

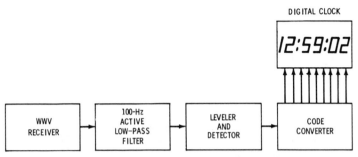

Fig. 10-18. Self-resetting, always-accurate clock.

The timing information of radio station WWV appears as a 100-Hz subcarrier on the audio. The audio output of a communications receiver (Fig. 10-18) is routed through an active filter that sharply attenuates everything about 100 Hz and mildly attenuates the lower frequencies to eliminate power-supply hum and extra noise. Since extra attenuation of 100-Hz signals can be contributed by the output amplifier and particularly by the output transformer, the signal is best obtained directly at the detector or at a low-level audio stage.

The output of the active filter is stabilized in amplitude and then converted into a series of pulses whose time width establishes one bit at a time of a complete code. The bits arrive at a one-per-second rate. A binary zero lasts 0.2 second; a "1" lasts 0.5 second, and a control or frame pulse lasts 0.8 second. The serial code is then converted to a parallel word with a shift register and then loaded into the timing system for automatic correction. More information on these techniques appears in *NBS Special Publication* No. 236.

PSYCHEDELIC LIGHTING

Most psychedelic lighting systems relate a visual display of some sort to music. One approach is shown in Fig. 10-19. We take an audio signal from the speaker system, chop it up into spectral chunks with a group of active bandpass filters, and then control a semiconductor controlled rectifier (SCR) or a triac in proportion to the amplitude of the signals in each channel. The SCR or triac then drives the load, several hundred to a few thousand watts of light. The blue lamps follow the lows, the yellows the accompaniment, the reds the rhythm, and so on, or whatever color combination is selected. The lamps can be projected onto a display or viewed through patterned but transparent plastic materials to create the final display effects.

Two-pole bandpass filters one-octave wide (2:1 frequency) with a 1-dB dip in them are a good approach (see example of Figs. 5-18

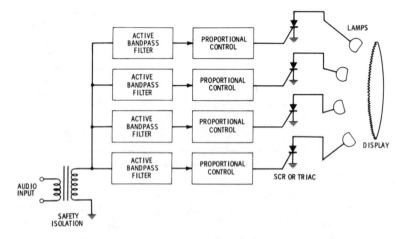

Fig. 10-19. Psychedelic lighting system.

and 7-16). A Q of 3.2 does the job, allowing any of the single op-amp circuits of Chapter 7 to handle the task. One important precaution in active psychedelic filtering is to leave gaping holes in the spectrum between the filters, perhaps as in the response of Fig. 10-20. This is important because of the strong harmonic content of music and the fact that you do not want all the lamps lit at all times. Further, the dropouts that happen when an instrument hits the guard bands between filters add considerably to the liveliness and interest of the display. Long-term agc or automatic volume control can be used to minimize long-term effects of loudness changes.

Design details on an early all-active-filter, six-channel, stereo psychedelic lighting control appear in the September 1969 *Popular Electronics.*

TONE SIGNALING

There are lots of different ways to communicate by means of tones over a wire, cassette recorder, or phone line. Some older and some-

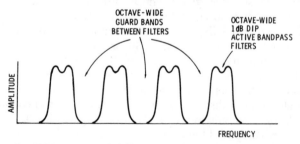

Fig. 10-20. Recommended filter spacing for psychedelic lighting.

what standardized systems used a single tone of one frequency, assigned to one of a dozen or more channels. The tones can be used in alarm systems, for industry telemetry, for process monitoring, or for anything else that needs a simple "yes-no" or "on-off" remote control. Active filters can be used with feedback for tone generation and can be used for detection or demodulation. Often the best detection route combines an active filter with a phase-lock-loop demodulator integrated circuit.

More modern systems use the standard touch-tone signaling frequencies. Typical system architecture is shown in Fig. 10-21, and the standard frequencies appear in Fig. 10-22. The touch-tone frequencies consist of eight assignments in two groups of four each. To signal, two tones are simultaneously sent, one tone from the low group and one tone from the high group. At the receiving end, both tones are detected and an output is provided only if both tones are simultaneously present. The two-tone system gives lots of immunity from noise and interference. As a side benefit of the system, the touch-tone dial of the telephone itself can often serve in place of a transmitter.

Other tone-signaling schemes that involve active filters appear in tape/slide synchronization systems and in cassette automatic phone-answering systems. The active filters normally serve to reject unwanted signals and false alarms while passing on the desired signal or tone.

MODEMS

A *modem* is a modulator-demodulator that lets you transmit digital data over a phone line or to a cassette recorder. A typical modem setup is shown in Fig. 10-23, and the key frequencies of two popular modem systems are shown in Fig. 10-24.

Active filters often greatly simplify modems designs. For modem transmitters, the active filter makes sure that only sine waves of the correct frequency are transmitted, eliminating any harmonic problems that could add noise and errors to data systems. At the receiving end, active filters are used to eliminate potential interference from out-of-band signals. In some circuits, they form the actual tone-detection circuitry. A more popular combination combines an active prefilter with a phase-locked-loop detector that does the actual tone reception.

Active filters for modems must be designed so that all frequency components of a digital pulse waveform are delayed by an equal amount. Otherwise, portions of the filtered signal will "slop over" into the available time slot for the next bit of information. This is called the *group delay distortion* problem. All active filters used in modem receivers must be carefully designed, or else they have to be special versions that carefully control the group delay to acceptable error-rate levels. More details on modem filters appear in Motorola Application Note AN-731.

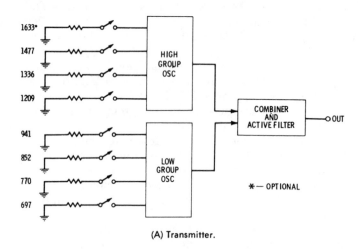

(A) Transmitter.

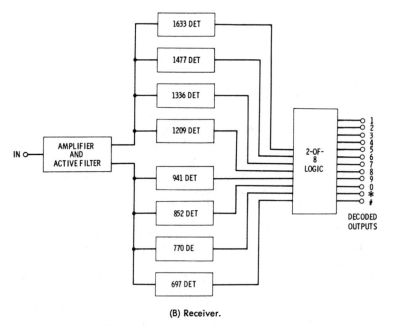

(B) Receiver.

Fig. 10-21. Touch-tone communication uses 2-of-8 code.

DIG	LOW GROUP, HERTZ				HIGH GROUP, HERTZ			
	697	770	852	941	1209	1336	1477	1633
1	●				●			
2	●					●		
3	●						●	
4		●			●			
5		●				●		
6		●					●	
7			●		●			
8			●			●		
9			●				●	
0				●		●		
*				●	●			
#				●			●	
SPARE	●							●
SPARE		●						●
SPARE			●					●
SPARE				●				●

Fig. 10-22. Standard touch-tone frequencies. Two tones are always sounded simultaneously.

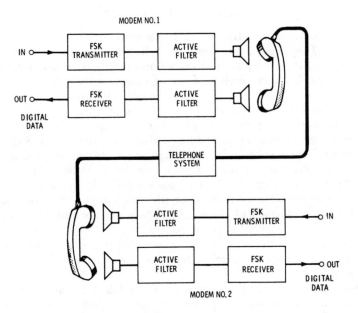

Fig. 10-23. Modems allow digital data transmission over the telephone system.

300 Bits/Second	**(Bell 103 Standard)**
	Two-Way (full duplex), Voice-Grade Line

Originate Mode: "Space" = "0" = 1070 Hz
"Mark" = "1" = 1270 Hz

Answer Mode: "Space" = "0" = 2025 Hz
"Mark" = "1" = 2225 Hz
(Also disables echo suppressors
on phone network)

1200 Bits/Second	**(Bell 202 Standard)**
	One-Way, Voice-Grade Line

"Space" = "0" = 1200 Hz

"Mark" = "1" = 2200 Hz

Fig. 10-24. Standard frequencies for two popular modem systems.

OTHER APPLICATIONS

The telephone industry probably uses more active filters than anyone else. They were the inventors of active filters, and they continue to lead in the development of theory and practical circuits for their use. Active filters are used to combine many voice or data channels together onto a common carrier for cable, microwave, or satellite transmission. Different types of active filters are used to continuously measure and adjust the characteristics of telephone lines for best quality and minimum error rate in digital communications. Some of these circuits are called *adaptive equalizers*. Echo suppressors, which improve voice communications over long distances but which must be disabled for digital data use, are controllable by active filters. Another application area involves the combination of supervisory and control signals with voice and data messages and, later, their separation for recovery of the messages.

We can expect to see active filters emerging strongly in automotive electronics, pollution control systems, and environmental sciences in general.

Geology and the earth sciences use active filters in studies of gravity anomalies, earthquake prediction, and studies of geomagnetic micropulsations. Active filters also aid the study of magnetic fields, particularly for proton magnetometers, where miniscule signal ranging from 1600 to 200 Hz must be extracted and accurately measured from a substantial noise background.

Much of medical electronics can use active filters to advantage, where tiny signals, be they heart rhythms, pressure signals, nerve impulses or whatever, are buried in a noisy environment and must be recovered for display, control, or analysis.

As the simplicity and ease of applying active filters becomes better known, we can expect more and more in the way of applications, particularly when the simple, compact, and low-cost methods we have shown you are used. What can *you* do with them?

References

Colin, Denis P. "Electrical Design and Musical Applications of an Unconditionally Stable Combination Voltage Controlled Filter/Resonator." *Journal of the Audio Engineering Society,* Vol. 19, No. 11, December 1971, pp. 923-8.

Fleischer, P. E. "Design Formulas for Biquad Active Filters Using Three Operational Amplifiers." *Proceedings of the IEEE,* May 1973, pp. 662-3.

Herrington, D. E., and Meacham, Stanley. *Handbook of Electronic Tables and Formulas.* Indianapolis: Howard W. Sams & Co., Inc., 1959.

Kerwin, W. J. "State Variable Synthesis for Insensitive Integrated Circuit Transfer Functions." *IEEE Journal of Solid State Circuits,* Vol. SC-2 September 1967, pp 87-92.

Mitra, S. K. *Active Inductorless Filters.* New York: IEEE Press, 1971.

Sallen, R. P. "A Practical Method of Designing RC Active Filters," *IRE Transactions Circuit Theory,* Vol. CT-2, March 1955, pp. 74-85.

Tobey, G. E.; Graeme, J. G.; and Huelsman, L. P., eds. *Operational Amplifiers— Design and Applications.* New York: McGraw Hill, 1971.

Weinberg, L. *Network Analysis and Synthesis.* New York: McGraw Hill, 1962.

Westman, H. P. *Reference Data for Radio Engineers.* Indianapolis: Howard W. Sams & Co., Inc., 1968.

Wittlinger, H. A. *Applications of the CA3080 High Performance Operational Transconductance Amplifiers.* Application Note ICAN6668, New Jersey: RCA, 1973.

Index

A

Ac-coupled low-pass filter, 118-119
Active filter
  advantages of, 8-9
  circuit, 10-13
  definition of, 8
  frequency range, 9
  Q of, 9-10
  types, 13-18
Adaptive equalizers, 232
All-pass response, 63
Amplifier
  current-summing, 23, 27-29
  voltage, 26-27
Amplitude response
  first-order
    high-pass section, 53
    low-pass section, 50
  second-order high-pass section, 62
"a," staggering factor, 102
Audio equalizers, 218-219
Automotive op amp, 42

B

Bandpass
  circuit
    Sallen-Key, 154-155
    variable-gain
      state-variable, 161
      tuning of, 162
  filter, 7
    biquad, 159, 161, 163
    circuits, 149
      multiple-feedback, 150-154
    design examples, 166
    design rules, 165
    multiple-feedback
      math analysis, 151-152
      tuning of, 153-154
    shapes, 91-92
    second-order, 96-100

Bandpass—cont
  response
    three-pole
      maximally flat filter, 114
      maximum-peakedness filter, 113
      1-dB dips filter, 114
      2-dB dips filter, 115
      3-dB dips filter, 115
      sixth-order filter, 109-111
    two-pole, fourth-order, 100, 102
  section, second-order, 57-61
Bandstop response, 63
Bandwidth, 92
  fractional, 92
  percentage, 92, 93
Bessel filter, 72
Best-time-delay filter, 72
Biquad, 34
  bandpass filter, 159, 161, 163
Brainwave research, 212-213
Butterworth filter, 72
  overshoot, 80

C

Capacitance values for frequency scal-
    ing, 145, 190
Capacitors
  disc ceramic, 194
  electrolytic, 194-195
  Mylar, 194
  polystyrene, 194
Carbon-film resistors, 195
Cascaded pole, 94
  synthesis, 91
Cascading, 18
Cauer filter, 63, 208
Center frequency, 92
Chebyshev filter, 72, 73, 77
Circuits
  bandpass filter, 149
    multiple-feedback, 150-154

1952-20
22-37